NUMBER 607

THE ENGLISH EXPERIENCE

ITS RECORD IN EARLY PRINTED BOOKS
PUBLISHED IN FACSIMILE

THOMAS HILL

THE SCHOOLE OF SKIL

LONDON, 1599

DA CAPO PRESS
THEATRVM ORBIS TERRARVM LTD.
AMSTERDAM 1973 NEW YORK

The publishers acknowledge their gratitude to
the Curators of the Bodleian Library, Oxford
for their permission to reproduce the Library's
copy, Shelfmark: 4°.D.30 Art.(3), and to the
Trustees of the British Museum for their
permission to reproduce pages E_6r,v, L_2r,v,
L_3v, L_4r, M_3r, M_4v, M_6v, R_7r and S_3r from
the Library's copy, Shelfmark: 532.L.4(1)

S.T.C.No. 13502

Collation: $A^4(^{-1})$, $B-S^8$

Published in 1973 by

Theatrum Orbis Terrarum Ltd.,
O.Z. Voorburgwal 85, Amsterdam

&

Da Capo Press Inc.
- a subsidiary of Plenum Publishing Corporation -
277 West 17th Street, New York N.Y. 1011
Printed in the Netherlands
ISBN 90 221 0607 1
Library of Congress Catalog Card Number:
73-7083

THE SCHOOLE OF SKIL:

Containing two Bookes:

The firſt, of the Sphere, of heauen, of the Starres, of their Orbes, and of the Earth, &c.

The ſecond, of the Sphericall Elements, of the celeſtiall Circles, and of their vſes, &c.

Orderly ſet forth according to Art, with apt Figures and proportions in their proper places, by
Tho. Hill.

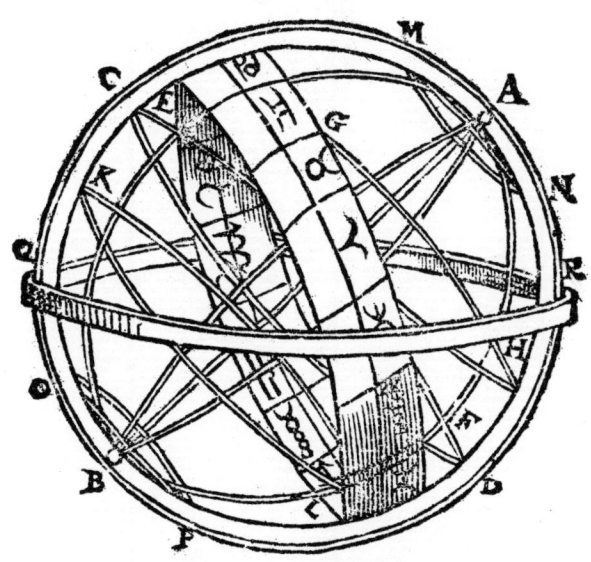

AT LONDON.
Printed by T. Iudſon, for W. Iaggard.
1599.

To the Reader.

DIuers haue writtē of sundry matters in former Ages, to the intent to benefit these our later times, wherin a man can name no kind of Art or Science, liberall or mechanicall, but there are as rare wits to bee found as euer liued since lerning florished. The reason is good that it should be so. For first, we haue come to our handes, vse and iudgement, whatsoeuer either antique or moderne Authors haue left behinde. Secondly, the gouernment (God be blessed) hath a long time (now these 40. yeares)

To the Reader.

bin so peaceable, that Students had neuer more libertie to looke into learning of any profession, for the inlarging of their vnderstanding. Lastly, the meanes otherwise, aswell out of the vniuersities, as in them, haue been and are so many and so good, to attaine to all knowledge, that I dare be bold to say, England *may compare with any Nation for number of lerned men, and for variety in professions. Of late a man of good merit, named* Tho. Hill, *painful with his pen whiles he liued, (as the world can witnesse, being possessed of sundry his works in Print) now deceased, left this Treatise Mathematical, intituled* The Rudiments of the Sphere: *which being found by iudgment of the lerned in the like profession, worthy the publishing, I haue, not only for the memory of the Author, but also for the profit of al wel*

as-

To the Reader.

affected Students, vndertaken to set forth. Wherein whosoeuer bestoweth time & labor to read, with a temperate and sober spirit, I doubt not but they shall be satisfied in all such points, as this Mathematician pretendeth to handle. His stile is not to be plausible, considering the subiect or matter whereof he discourseth, doth restraine him, both to tearmes of Art, and phrases consonant. But for his order and facility (such as a profession of this nature will beare) better to be conceiued, than some (none dispraised) that haue written of the like argument. It is not vnlike, but he would (if God had spared him longer life) haue held on as he began, to set forth for the common good of his and our Country diuers necessary works. This seeming his last, and whereof there is vse both on Sea and Lande, printed according to his

owne

To the Reader.

owne copy, and the Figures anſwerable to the patterns as they were drawne by pen, now newly ſtept into the world, receiue & read friendly, find no fault, but accept the good minde of the deceaſed man, and thanke them by whoſe meanes this booke, which otherwiſe might haue beene loſt, is ſet on foot, and come abroad. April. 8. 1599.

Yours W. I.

THE FIRST PART OF THE RVDIMENTS OF THE SPHERE OF HEAVEN,
of the Starres, of the Orbes of the Starres, and the EARTH.

Eing this litle Book of the Sphere dooth intreat of that part of Astronomy, which sheweth the diuers motions of the Celestiall Orbes and starres, the magnitudes and distances of their bodies from the Earth, with all the diuersities and néerenesse of appearaunces in the Planets, and fixed starres: therefore doth the Author write of the Principles of the same, in this Treatise of the Sphere, to the great commodity of many young Students in the Art. For this containeth onely the intreating of the Sphere: that is, of a perfect and very round body, containing diuers Circles, which the learned doo also call a Materiall Sphere: of the Celestiall appearances that it describeth in the Instrument, named of them the Materiall Sphere.

B i. Now

The first Part

Now this teacheth fiue definitions of the same: twoe of the Sphere, one of the Center, one of the Exe tree, and one of the Poles of the World.

1. What a Sphere is.

EVCLIDE in his eleuenth Booke, thus defineth a Sphere. A Sphere (which in Latine is a Globe) saith he, is a sound Figure, made by the turning of a halfe Circle, the Diameter of which halfe Circle continuing so long steddy, vntill it bee brought again vnto the place, wher that Figure began to be drawne. Or thus: A Sphere is such a round and sounde body, which is described by the drawing about of the halfe Circle.

Theodosius teacheth another definition of the same: That the Sphere is a certaine massy Body, or sound Figure, inclosed with an vpper face or platforme, in whose middle is a Pricke, from which all lines drawn from the Circumference or platforme are equally distant one from the other: and this Pricke, of him named the Center of the Sphere; and like of the Globe.

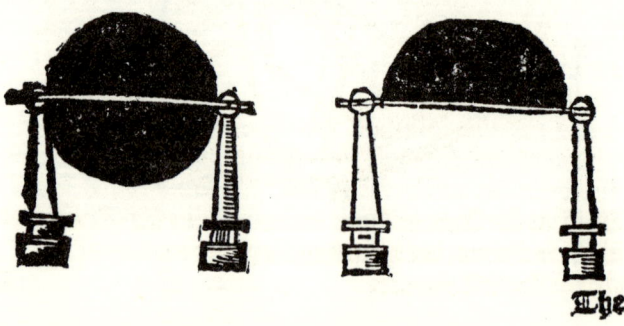

The

of the Sphere.

The halfe Sphere is contained in the halfe of the Globe, and greatest Circle of the Sphere.

The Ere-trée of the Sphere is a right line, about which the Sphere fired, doth the halfe Circle drawne about describe.

There are also two maner of Orbes, as the Solyde Orbe, and hollow Orbe: the Solyde is named the Globe or Sphere, which only containeth one round vpper face, and the same imbossed hollow outward: but the hollow Orbe differeth, in that the same hath two vpper faces, the one imbossed outward, and the other hollow within. Also the Orbes of all the fired starres, and Planets, are like hollow, and not Solyde.

A materiall Sphere, is that which is made of ringes, or Circles, in such a manner framed, deuided, and dispo-

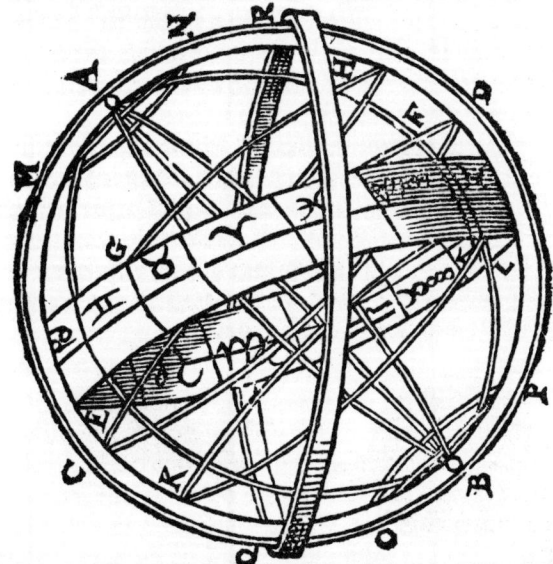

sed, that the same in some maner may expresse and shew forth to the eie, both the standing and motion of the Circles in the first mouer.

To the Sphere belong these differences, A Pricke, a Line, Dyametre, Parallels, an vpper face, a Center, Exe-trée, and Poles.

A Pricke or note is of no bignesse, but the beginning of Magnitudes, which in the order of nature goeth before them, and not made as a part: in that a Line is not made of Prickes, nor Prickes are the partes of a Line. For if an infinite number of Prickes were heaped and ioyned togither, yet woulde those neuer make a Line: so that a Line is caused, through the drawing of a Pricke into length.

A Line, is a length without bredth and déepnesse, and ended with two prickes, which cannot be comprehended, but by Imagination.

A Dyametre, is any right line drawne by the Center of a Circle, and middle of a Figure: whether the same be plaine, Solyde, round, or cornered: whose endes reache and come from side to side of the Circle about, and deuideth the Circle into two equall partes or iust halfes.

The Parallels are two right Lines equally drawne, which extended on any Platforme vnto an infinit length, doo alwaies kéepe one like distance, and neither draw néerer, nor touch togither.

An vpper face, is a length and breadth without déepenesse, made by the drawing of a Line into breadth: Of which, the plaine vpper face is that, which is expressed with those straight Lines which it hath: that neither the middle riseth vppe, or is raised at the endes, nor the same falleth within. The Sphericall vpper face is distinguished, into an imbossed and hollow vpper face. The imbossed, is the outward compasse about of the Sphere, or bodies round: but the hollow vpper face, is the inwarde compasse about in the hollow Orbe, or the bodies hollow.

A Center, is the middle Prick in a Circle, from which all right Lines drawne vnto the compasse about the same,

are

of the Sphere.

are equall betwéene the one and the other. Also a Centre of the Sphere, is a middle Pricke in the Sphere, from which all right lines drawn vnto the imbossed vpper face, doe agrée in length.

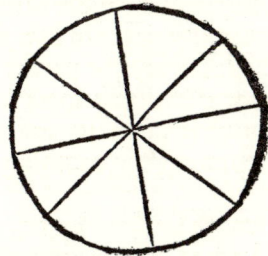

The Exe trée, is a right line drawne by the Centre of the Sphere, and with both his endes pearcing vnto the sides of the imbossed vpper face, about which the Sphere is turned.

The Poles, are the verie endes of the Exe-trée, appearing on each side, about which heauen is turned. Also the Poles of the turning, are named the endes and tops, and named besides the ending pointes of the Exe-trée, drawne by the Centre of the Sphere: about which the Spher and Circles of the Sphere are turned.

The Poles of the Sphere, and Circles described in the Sphere, are pointes consisting in the vpper face of the Sphere, from which all right lynes drawne vnto the compasse of the Circles are equall.

The first Part

Although euery greater Circle in the Sphere of the worlde hath his Poles, yet is oftner mention made of the Poles of the worlde, or Equinoctiall of the Zodiacke, and Horyzont.

The Poles of the world or Equinoctiall, are the two extreame poyntes and both endings of the Exe tree of the world, about which the Sphere is turned.

The one of the Poles, which eleuated sheweth to vs dwelling into the North, and turneth alwaies about in our sight, is named the Boreall and North Pole, of the litle Beare neare to it: which Pole is continualy seene and knowne by the two starres neare to it; of which the one is notable and cleare, of the thirde bignesse, standing at the ende or top of the tayle of the little Beare, distant from the very Pole 4. degrees, and 9. minutes. But the other dimmer, of the fourth bignesse, not farre distant from the other star afore, is come nearer, and doeth scarcely differ nowe 50. minutes from the place of the true and very Pole.

If any will know the Pole of the world or Pole starre, let him turne his face towarde the North, the sky then being cleare; leauing the East on the right side, and west on the left side, and he shal see in the little Beare seauen stars, placed like to the forme of the starres of the great Beare, which are brighter. Of these stars three doe fashion the tayle, and that which is in the top of the tayle, is named the Pole star, which declineth in our time from the Equatour 85 degrees, and 53. minutes. So that beeing no further off by this declination from 90 degraes, the distance of it from the very Pole shall remaine and bee 4. degrees, and 7. minutes: and this starre also in processe of time shall be ioyned with the Pole.

The other Pole, which through the swelling of the earth, is continually hid to vs of the North, is named the Meridionall or South Pole, right against as it were the North Pole. Beeing euermore so lowe depressed, as the

North

of the Sphere.

North Pole raysed, in any countrey to the North, aboue the Horizont.

The Poles of the Zodiacke, are continually so farre distant from the Poles of the world, as is the greatest declination of the Sun, which in our time is founde and noted by obseruations, to be of 23. degrées, 28. minutes, and 30. seconds. But the Boreall or North Pole of the Zodiacke, distant from the two dimme starres, in the tryangle of the Dragon (being stars of the sirt bignesse) which a right line by imagination drawne from the third starre in the tayle of the little Beare, vnto that constellation named Lyra, expresseth the same, that it is but a little further distance then two degrées.

The Poles of the Horizont, are the twoe extreame pointes or ends of the right lyne, drawne out of the Center of the Earth, by the top ouer our head, vnto the opposit places of the Dyametre of the Merydyan: of which the one, directly ouer our heade, named the verticall poynte, and of the Arabians Zenith, the other right against, named of them Nadir.

And the Poles of each of the greater Cyrcles, doe differ from their Circles 90. degrées, or a quarter of an other greate Circle of the Sphere. For by the 23. proposition of Theodosius of the Sphere, a right lyne drawne from the Pole of his Circle vnto the circumference, is equall to ech of the foure quarter sides, descrybed within the same Circle, which foure quarter sides, doe deuide the Circle into foure quarters.

Yet are each greater Circles of a Sphere, equall betwéene the one and the other. But séeing the equall right lynes, doe take away the equall compasses of the equall Circles, therefore should a lyne drawn from the Pole vnto the circumference of his Circle, take away of the greater Circle, one quarter of the other Circle drawne by the Pole, euen like as the sides of a quarter described within

A iiij.   the he Circle. By which appeareth, that the Poles of the greater Circles, doe differ or be diſtant by a quarter from their Circles, as is aboue written.

What the World is, and into how many
partes the same is deuided, with the motion of the celeſtiall Orbes.

He Worlde after Orontius, is defined to bee a perfitte and an entire compoſition of all things: a deuine worke, but finite, and continually to be merueled at: adorned with all kindes of formes and ſhapes of bodies that nature coulde make, which in all partes are procreated and appeare: and thoſe firſt created by God (ſo well in Earth as in Heauen) by his onely word of nothing, to th'end the ſame might bée a proper manſion place for man, in which he might dayly behold, and make knowne.

Ariſtotle teacheth two definitions of the world: the one, that it is an apt frame wrought, conſiſting both of heauen and earth, and of the celeſtiall and inferiour bodies aptly diſtributed, and of other naturall thinges in them contayned.

The other definition is, that the world is a perfit body, and moſt perfit rounde forme, contayning the ordinance and diſtribution of bodies, created by God to tend vnto a purpoſe, which by God, and through God is preſerued.

The parts or Regions of the worlde, are two: as the Æthereall, and Elementary.

The Etheriall region is the higher and vpper parte of the world, which encloſeth the Elementary region, being wholy cleare and the light perfitte, and contayning the

Orbs

of the Sphere.   9

Orbes of all the fixed starres and Planets, distinguished by a certaine order, free of all mixture and all strange qualities, nor harmed by any alterations. In which the celestiall bodies are drawne about by certaine and continuall orders and times of the motions, that they may so cause the diuersities of times, dayes, yeares, and moneths: and as well in the Elementary nature, by his motion and light, ingender, mixe, and temper togither the first qualities, and prepare also other effects.

The Elementary region, is the nether part of the world, which is contayned within the hollowe vpper face of the Moones Orbe and Sphere, in which are all corruptible bodies, and thinges harmed by diuerse alterations, except the minde of man: the causes of which, are the contrary actions of the first qualities. Also the foure Elements are simple bodies, which into parts of diuers formes cannot be deuided: yet through the mutuall commixion of these, are diuerse kinds of bodies caused. Therefore, whatsoeuer bodies are in the Elementary region, bee either simple, myxt, or compound bodies: In that the mixt bodies, are all those which may be deuided into parts of diuerse kinds.

To these of the foure Elements, the next ioyning within the hollow vpper face of the Moones Orbe, is the most thinne Ayre (being the lightest of the Elements) kindled, through the dayly mouing about of the celestiall circles: which for this congruency with the fire (named the elemental fire) that is dayly drawn about by the Orbs compassing it, which may appeare by the Cometies, and other fiery kindes, ingendred in the same Element of a hot and dry vapoure, that are likewise caried about.

The next within that doth the ayre runne, being a heauier Element then the fire, yet lighter then the water: which also is drawne about by a like motion, as may appeare by the clouds, and other like impressions ingendred in the same, but to the nether region of the same, consist the late-

laterall motions, as wee dayly see by the blowing of the windes. Further, Vitellio in his tenth booke and 60. chapter affirmeth, that the cloudes are distant from the vpperface of the earth 25000. paces, or 13. Germayne myles. But according to some writers, they are vnequally distant from the earth: as somewhiles further off, and somewhiles neerer to the earth. For when the cloudes are furthest distant from the earth, they are but 772000. paces, and being nearest the earth, are 288000. paces distant. To conclude, this Elemente compasseth and encloseth, both the earth and water by his largenesse.

The nexte Element to the Ayre which mooueth, is the water, for the same is moued by a motion of flowing and ebbing, which it maketh after the motion of the moone; in that it floweth sire houres, and ebbeth so many, vntill the moone by the motion of the first mouer, hath passed about all the quarters of heauen. Also the water hath a motion, and that downward into the earth, so that these two ioyntly annexed, make as it were one body. Yet the earth beeing the heauiest Elemente, hath a motion attributed as it were simply downwarde vnto the middle: notwithstanding agrees of all men, that the same is immouable, and the Centre of the world.

These foure: that is, the fire, ayre, water, and earth, are named to be the foure Elements, and both the simple, and Original matters, of which all mixt bodies are compounded and made.

The proofe that there is onely fowre Elements, is this: that to each Element the two first qualities agree, and the Combynations the like of the foure qualities: as of heate and dryeth, which consist in the fire: of moysture and heat, which rest in the ayre: of coldnesse and moysture, which be in the water: of dryneße & coldnesse, which is found in the earth.

By these it is euident, that there are but foure Elements,

of the Sphere.

ments: of which heate excædeth in the fire, moysture in the ayre, coldnesse in the water, and drynesse in the earth. To conclude, it appeareth, that heat with colde, and moysture with dryeth, cannot aptly be ioyned.

What the Starres are, and that, as to the motion of their Orbes, they are carried about.

The Ethereall region contayneth the Starres, which are the thicker parts of their Orbs; perfit rounde, cleare, most pure and simple, and frææ of any mixture, except the Moon which is darker then the others, yea variable and shadowed. And these fastened to their Orbes, by which in certayne continuall and appoynted times and orders, are drawne about, and performe their returnes in the determinate spaces of times, and those continually agrææing in themselues, that they may so ingender the differences and orders of times, and in the inferiour nature prepare and cause the first qualities, and other effects.

The Sunne the fountaine of light, doth not onely giue light and make shine cleare the inferiour bodies, but the superiour also, by the brightnesse and light of his beames.

But the Stars sææing with a borrowed light they shine, which is far weaker then the sunnes, therefore with that strange light which they take properly of the sun doe they shine, although vnlike to the sun.

For into all the starres, which by nature are rounde about thynne, and penetrable, is the sunnes light equally shed and pearceth, and so filleth all, that they are subiect to no times of encreasing and decreasing of light.

But

The first Part

But the Moone, seeing it is an vnperfit body, and that it hath the partes some where thynne, & somewhere thicker and better compact: therfore doth it not equally, nor round about receiue the sunnes light. So that the thynner parts take more of the sunnes light, and of the same doe clearer shine. But the lesser shadowed parts which also are seene, appeare darker, as the spots in the moone do shew.

That the bodies of the starres are round, doe the round formes in the Eclipses of the sun and moone shew: yea in what parts of the world those Eclipses happen, doe the bodies also of the starres at that time appeare perfit round: Although the bodies of the starres be knowne (by sundry reasons) to be round as a bowl, yet by their great distance from the earth, appeare to vs as playne or flat.

Nor the Starres are not moued by their owne proper motions, but by the Accydentary, as vnto the motion of the Orbs, to which they hang, as partes vnto the motion of the whole. For to euery round body doe two proper motions onely belong, as a mouing to and fro, and turning about. Therefore the Starres (seeing they be round) are by some proper and principall motion caried round. But the fixed Stars are not so moued rounde, in that they turned about, doe not altar the same face or body which they once turned and shewed to vs: but that the same shoulde of necessity happen like, being turned round in one place, about their Exe-tree, with the others in the same motion bæing in the parts far distant, and the others then set and hidde vnder the earth. Nor are they turned hither and thither, in that they neuer change the standing and place which they haue in their Orbe, which to those caried hither and thither woulde happen. Therefore, not by a proper and chiefe motion are they caried about, but by an accidentary drawing about of their Orbes, which what the same is, shall after appeare.

That

That Heauen is drawne round.

The Ethereal region, do the Philosophers also name *quinta essentia*, or as it were a fifte body, constituted aboue the foure Elementes, being incorruptible & deuine, consisting of the noblest and purest part of the ayre. Which also is placed aboue the holow vpper face of the moones Orbe, that reacheth vnto the holow vpper face of the highest heauen, being most pure, perfitte rounde, continually caried about, and bright appearing.

This parte of the world being the Etheriall region, is named heauen, which alwaies drawne about by a meruelous swiftnesse, is deuided into nine Orbes or Spheres. Although sundry Astronomers, as Alphonsus, Iohannes de monte regio, Purbachius, and others, haue added a tenth Sphere, through the third contrary motion founde in the eighth Sphere, named of Thebit benchore (the first inuenter of the same (*Motus trepidationis*, or the going and comming of the eight Sphere.

The first and vppermost Orbe, is named the first mouer. The second is that, which is named the ninth Sphere or Christaline heauen, but of Ptholomy, named the firmament or Orbe of the fixed starres. And the thirde is that, which (of them) named the eight Sphere, ouely added through that motion of the trembling, or as it were a mouing foorth and returne of that eight Sphere; which properly is caused, in the two small Circles about the heades or beginnings of Aries and Libra: through which diuerse motion of the eight Sphere, do the Equinoctials and Solstices

stices come and beginne sooner by certaine daies, and the suns greatest declination deminished (and dayly doeth) to that in Ptholomie and Hipparcus time, which then was 23. degrées and 52. minutes, and 30. seconds. And for these haue Alphonsus and sundry others, attributed diuers motions to the eight Sphere, adding a ninth and tenth Sphere to it.

That there are but eight celestiall Orbs which may be seene.

Although Ptholomy affirmeth, that there are nyne Orbs equally distant, yet are there but eight which may perfectly be séene and decerned with the eie, both in the standing, variety of motions, and differing in the periodes or courses. Also they are in such order disposed, that no Orbe hindereth the motion of another néere to it. As the Sphere of the fired stars, and the seauen Orbs of the Planets. And most certaine it is, that some of the fired stars are drawn by a swifter motion, and others by a slower motion, and that the Apogea or ascentions also of the Planets are changed, after the order of the signes.

The Orbs of the Planets thus containe and compasse one an other, as first the Sphere of Saturne being nighest the firmament (of which being compassed) doeth like containe Iupiters sphere, and Iupiters, doth in the same maner inclose Marses sphere, and Marses in like order, the sunnes sphere, nexte the suns, doeth containe Venus sphere, which like doth compasse, Mercuries sphere, and Mercuries doeth containe the Moones sphere, being the lowest and smallest sphere. And euery of these spheres, hath a star a péece, named

of the Sphere.

med eratical stars or planets: which stars haue euery one their proper Orbe seueral, his motion seueral, and vnlike in time one to another, in that they appeare one whiles neere togither, and another whiles are seene far distant asunder.

By which it agreeth, that their equall motions, to appeare to vs vnequall, either through the Poles of the Circles, diuers from the Poles of the worlde about which they bee turned, as are the Poles of the Zodiacke, vnder which the eratical stars are continually drawn and moue; rather for that the earth is not the Center of those Orbes, by which the Planets are caried and moued about.

So that when we consider those mouings by the Center of the worlde, then is caused, that they seeme to vs as they were encreased in a greater bignesse, when as we beholde and see them neere hand, and that lesser in bignesse, when we see them placed far off. Euen so in the equal circumferences of the Orbes, through the diuers distance of sight, wee like obserue the vnequall motions, by the equall times.

Yet indeede neyther of these happeneth, but that they are drawne aboute by vnchangeable spaces, beeing a like distant, and keeping one manner of bignesse. For if this were, then the sun, or any other star being in the middle of heauen, should seeme or appeare bigger (which it doth not) then being in the East, or West part.

And the contrary we sundry times see, when as the sun or any other star, appeareth bigger in the quarters of the East, and West; which is not caused by reason of the shorter distance, but for that his beames in the vapors, which doe thicke ascend (both in the winter time, and in raynie weather) that hang in the ayre betweene our sight, and the body of the star, are then broken: which breaking of them, doth cause the star to appeare far bigger to the eie, then in deede the same is.

And

16   *The firſt Part*

And that a readier and easier knowledge may bee had (after the mind of Ptholomie) of the firſt moouer, and celeſtiall Orbs with the number of the Circles and Elements incloſed within the firſt moouer, conceiue this figure here following moſt aptly drawn and set out for thy further inſtruction.

This Figure declareth the number, diſpoſition, and order of the celeſtiall Spheres, about the Globe of the Earth.

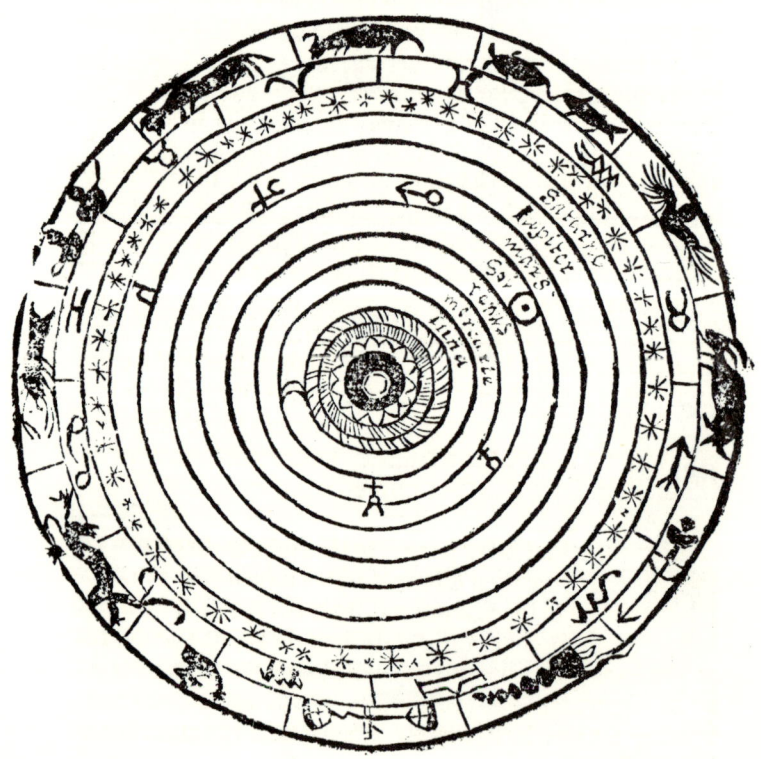

That

That there are two first motions of the celestiall Orbs.

Although the celestiall Orbs are seuerally drawne, by proper and vnlike motions: yet be there two first motions, that are manifest, both by obseruation, and iudgement of the eie. The one is of the first mouer, which Ptholomy attributeth to the ninth Sphere, that is onely drawn about by an equall swiftnesse, from the East into the west, and from thence againe into the East, vppon the Poles of the worlde or Equatour, in the space of a naturall day or 24. houres. And this first mouer draweth with it al the other Orbs, much like a ship, which being at full saile, doth drawe and cary al her men, & other liuing things which are in the Shippe. So that by this motion of the first moouer, the neather Orbes, which the first mouer compasseth, are drawn once euery naturall day, or in the space of a day and night, about the earth. Also this first mouer, doeth not onely describe and measure a naturall day, but causeth times, and diuersities of dayes and nights, with the proper motion of the sun; and it dayly bringeth vp stars to be sæne, and carieth vp to the highest, and after hideth them againe, vnder our Horizont in the west. Besides it is the common measure of al the other motions.

The other motion is proper to the eight Sphere, and to the Orbs of the seauen Planets, in the which they are contrarily caried to the first mouer, as from the west into the east, in mouing vnder the Zodiacke, and about the Poles of the same: and not in Parallels from the Equatour equally seperated; but are drawne much slower, yea and vnlike. As by a like example; when a ship by a most swift

C.i. course

course is caried into the west, yet may the Mariners and others in the ship walke forwarde in the meane time into the East: Euen so is this second motion of all the other Spheres vnder the Zodiack, vpon the Poles of the Eclipticke. Also by a swifter motion are they caried, and soner performe their courses, which are nærer to the earth, and contrariwise moue by a slower pace, and in longer time compasse and wander aboute the signes of the Zodiacke, which are further distant from the earth. Also in the middle of their courses (as it were) each doe often slacke or bee slow, and often times stay as vnmoueable, and sometimes are retrograde, after againe quicken their course, and by their swiftnes recouer that lost of the former tariance. So that they neuer kæpe one manner of way, but one whiles from the middle iourney of the Zodiacke doe wander into the North, and another whiles into the South. To conclude, they be ascended high from the earth, when they are named Apogei, and discended again vnto the earth, when they are named Perigei.

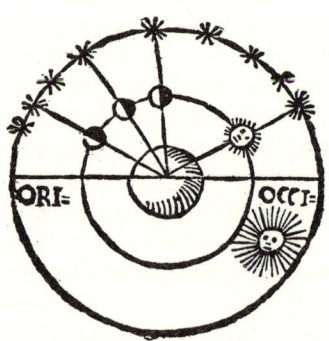

That there are two kindes of Starres, the fixed, and the Planets.

ALL the fixed stars that hang to the firmament (or as Ptholomie affirmeth) to the eight Sphere, are named fixed: not for that they consist vnmoueable, but that they moue so meruellous slow, that by iudgement of the eie they cannot be perceiued to moue: yet the practitioners haue and do find both by reason and obseruations, that they alwaies seperated by vnmoueable spaces one from the other, and are caried in parallels as it were in their Orbe.

Ptholomie, Aristillus, Timochares, with the obseruations of Hipparchus, ioyned vnto those which he knew, noted that the fixed stars in a hundreth yeeres moued one degree. But Copernicus being of later yeares (as about the yeare of Christ 1525.) that examined the obseruations of the auncient men, and compared them vnto those noted of him, founde that not so many as a hundreth yeares, but in seauentie and two yeares, that the fixed had gone one degree: and that in euery Egiptian yeare (which containeth 360. dayes) the fixed to haue moued 50. seconds, and in a day 8. thirdes, and tenne fourths. And so performed their whole course, in twentyfiue thousand, eight hundreth, and sixteene yeares, 25816.

Of these knowne and numbred of the ancient men, are 1022. which they haue deuided into sixe differences of magnitudes; and to these haue added certaine obscure, and certaine cloudy stars.

The fixed stars of the first bignesse, of which are numbred

bred to be fifteene, and that both in bignesse and brightnes exceede all the other starres, and in body exceede the earth 107. that is, a hundreth seauen times, with a eleuen threescore fourths.

The fired stars of the second bignesse, of which are accompted 45. do exceede the earth in greatnes almost eighty seauen times.

The fired stars of the third bignesse, of which are numbred to be 208. doe ouerpasse or exceede the earth seauenty two times, with a third part almost.

The fired stars of the fourth bignesse, of which are reckoned to be 474. that are fifty foure times greater then the Earth, with a halfe or a little more of the earth.

The fired stars of the fift bignesse, of which are noted to bee 216. that exceede by their greatnesse the earth 31. times.

The fired starres of the sirt bignesse, of which are noted to be 50. doe exceede the earth eightœne times, and a little more.

The darke stars, are accompted 3. in number: and the clowdy stars, are reckoned to be fiue.

The fired stars do differ in brightnesse, standing, color, twinckling, and especially in the configuration. Many of the fired stars also with the effects both in the bignesse and brightnesse, being notable and neere together, the ancient men that deuided them by a certaine reason, haue gathered, digested, and fashioned them, into forty and eight images or similitudes. And vnto them through the congruence and similitude of the noted formes or figures, they gaue apt names: and for the same cause especially, that they might the easier and sooner be discerned, knowne, and noted by their peculiar names. Also they deuided the stars, by the standing into the North and South, and the signes of the Zodiacke. The Northerly stars that decline from the Eccllipticke into the North, are twenty and one in num-

of the Sphere. 21

number. The Southerly, that decline from the Eclipticke into the South, are fifteene in number. The images that are named the signes, are twelue in number, which consist in the Zodiacke.

Of the celestiall Images, and of their diuers names, being in number, 48.

Of the Northerly.

1 He litle Beare hath seuen stars, and of those, that star which is in the top of the tayle, is named the Pole star.

2 The great Beare, hath 27 stars, 8 without forme.

3 The Draggon hath 31. stars.

4 The image named Cepheus, hath 12 stars.

5 The image Bootes or Lanceator, hath 22. stars.

6 The Boreall or Northerly crowne hath 8. stars.

7 The image kneeling or Hercules, hath 28. stars.

8 The Harpe or Griepe falling, hath 10. among these the brightest is that named the Harpe.

9 The wilde Swanne or Griepe flying, hath 17.

10 The image Cassiopia, hath 13. stars.

11 The image Perseus carying the heade Algoll, hath 26. stars: of which, those which are on the lefte hande doe make the head Algoll or Gorgons, 3. without forme.

12 The Cartare, hath 14. starres, among those, that which standeth on the lefte shoulder, is the Goate: and the other two are named the Kids.

13 The image Ophiulcus o₂ Serpentarius, hath 24. stars.
14 The Serpent, hath 11. stars.
15 The figure named the Arrow, hath 5. stars.
16 The Egle hath 9. stars, 6. without forme.
17 The Dolphin, hath 10. stars.
18 The deuision of the Horse, hath 4. stars.
19 The winged Horse o₂ Pegaius, hath 20. stars.
20 The image Andromeda, hath 23. stars.
21 The Tryangle, hath 4. stars.

        The Summe of all the Starres, are 360

The 12. Signes of the Zodiacke.

1 The image named Aries hath 13. stars.

2 The image Taurus hath 23. of these fiue in the fore-head of Taurus, named Succulæ o₂ Hiades, and the greatest star of Hiades in the Southerly eie, named Palliticium, and Pleiades on the back of Taurus, 11. without forme.

3 The images named Gemini, are 18. stars: of which Castor o₂ Appollo goeth before, Pollux o₂ Hercules, followeth.

4 The image Cancer hath 9. stars, among these Presepe, and the cloudie star in the Breast.

5 The image Leo hath 27. stars: of these, that which is in the heart of the Lion, named Regulus, 8. without forme: among which is that constellation, named Berenices bush of haire, betwæne the tops o₂ endes of the Lion and great Beare.

                                        6 The

of the Sphere.

6 The image Virgo hath 26. stars: of these, that which is in the right wing, especially Northerly, is named Vindenuator, but in her left hand a bright star, named the eare of corne, 6. without forme.

7 The image Libra, and klees of the Scorpion, haue 8. stars, and 9. without forme.

8 The image named the Scorpion, hath 21. stars, the midle star (of the three stars) placed on the Body, is named the heart of the Scorpion, and 3. without forme.

9 The image named Sagitarius, hath 31, stars.

10 The image named Capricornus, hath 28. stars.

11 The image named Aquarius, hath 24. stars, and 3. without forme.

12 The images named Pisces, haue 34. stars.

> The Summe of all the Starres, except Berenices bush, are 364.

Of the Southerly.

1 The image named Cœtus, hath 22. stars.

2 The image named Orion, hath 38. stars.

3 The image named the riuer, or Eridanus, or Potamos, hath 34. stars.

4 The image named the Hare, hath 12. stars.

5 The image named the Dog, hath 18. stars, of which that in the mouth, is named Alhabor, 12. without forme.

6 The image named the little Dog, or Caniculare star, hath 2. stars; of which the brightest is that named Proion or the Dog-starre.

7 The image named the Ship, hath 45. stars, of which a bright star going before in temone.

8 The image named the Water Serpent, hath 25. stars

and 2. without forme.

  9 The image named the Bucket or great Cup, hath 7. starres.

  10 The Rauen or Crow, hath 7. stars.

  11 The image named Centaurus, being one halfe like a man and the other halfe like a horse, hath 37. stars.

  12 The Beast which the Centaure doth holde, being a Wolfe, hath 19. stars.

  13 The image named the Aulter, hath 7. stars.

  14 The Southetly Crowne, hath 13. stars.

  15 The Southerly fish, hath 11. stars, and 6. without forme.

The Summe of all the starres, are 316.

The milkie way, which Ptholomie nameth Galaxian of the white and milkie colour, is a heape of most small stars, and dimme to sight; of which is a certaine confused gathering together, and abundance as it were encreased, that no seuerall light is decerned: and the same (in the maner of a girdle) compasseth and encloseth heauen about. The same also is vnequal, and differeth in the standing, latitude, haunt of stars, and in the colour very much. It is somewhere decerned clefte, but the parte going before, is neither whole, nor maketh a whole swathe or inclosure about, but lacketh about the swan and Aulter. And the part folowing whole, being in no place broken off with a space, and stretched thwartly in heauen: and from the partes of the Zodiacke Northerly, it passeth by Gemini, and Sowtherly by Sagitarius, and Capricornus.

Of

Of the Planets.

The Planets, named otherwise the erring and wandering starres; not for that they erre by a wandering and vncertaine motion, but in that they are caried aboute by a diuers and vnlike motion. For sometimes they goe foreward, and sometimes retrograde; sometimes are hidden and cleane out of sight, after they appear and shew themselues. Againe, they goe before, and follow the Sun. They are caried swift, and their motions againe so stayed, that they are moued in a maner nothing at all, but seeme as they were stayed for a time. From the sunnes way, one while caried into the South, and another while caried into the North, and then vnto the same way drawne backe againe: so that their iourneies being passed and finished, they steadyly repeate their old courses by the like order. Of these are seauen, and each caried in their proper Orbs, and compasse about the Zodiacke, in vnlike spaces of time,

SAturne highest of the Planets, and most slow in course, being cold and dry, pale to a leady colour, and perfourmeth his course in 30. yéeres, being ninty times, with an eight part greater then the earth. And the highest ascention or pointe of Saturnus Orbe (which at this day is in the 29. degrée of Sagitarius) is from the earth 20072. semidiametres, with a fourth part almost, and 15. minutes. But the lowest point of Saturns Orbe, is distant from the earth, 14378. with a third part, and 20. minutes.

Iupiter

IVpiter being next vnto Saturne, temperate, and so cleare or bright, that he giueth in a maner a shadow (especially when he is Perigeus or lowest discended to the earthward) and he compasseth about the Zodiacke in twelue yeares. But Iupiter giueth this proper shadowe, when neither the lights bée aboue the earth, nor Venus néere to him. Hée is greater then the earth, by ninty fiue times, and a half part almost. And the highest ascention of Iupiters Orbe, which possesseth the seauenth degrée of Libra, is from the earth 14369. with a fourth parte almost, and 15. minutes, but the lowest point of Iupiters Orbe, is from the earth distant 8853. semidiametres, with a ninth part and 45. minutes.

MArs béeing hot and dry, and shining with a fiery colour, doeth goe about the Zodiacke, in the space of two yéeres. He is named the fiery Planet, of his shining with a fiery colour, or of the effect which foloweth by him, in that he burneth and dryeth vp. He is one time greater then the earth, and a little more then a third parte. The highest ascention of Marses Orbe, that obtaineth the 28. degrée of Leo, is now distant (after Albategnius) from the earth 8022. semidiametres: but the lowest poynt from the earth, is 1176. semidiametres.

THe Sunne obtaineth the middle place betwéene the Planets, wholy and throughly bright, being the fountaine and Author of light: which by his motion expresseth and deuideth the spaces of the Zodiacke, and by his going about, haue the signes their names. He is greater then the earth (after Ptholomie) a hundreth thréescore and sixe times, with thrée eight parts. But after Capernicus, the sun excéedeth the earth, a hundreth thréescore & two times, with eight parts lesse. The highest ascention or poynte of the suns Sphere, which now possesseth the seuenth degrée of Cancer, is from the earth distant 1179. semidiametres,

but

but the lowest poynt of the sunnes Orbe, is from the earth distant 1065. semidiametres.

Venus next to the Sun, being cold and moist, white in colour, clearer and brighter shining then Iupiter, and is caried about (like the Sun) in a yeares space, and both goeth before and foloweth the sun; nor is further distant in the spring of the morning from him, then 46. degrees, and 47. minutes: but in the euening, shee is seene digressed from him, vnto 47. degrees, and 35. minutes. When shee goeth in the morning before the sun, shee is named the day star: but when shee followeth the sun in the euening, shee is then named the euening star. Lesser shee is then the earth, but her true quantity is yet vnknowne: for that some affirme her quantity to be the 28. part, and others the 37. part of the earth. The highest ascention of Venus Sphere, that obtaineth the 18. degree, & 20. minutes of Taurus, is from the earth after Albategnius 1070. semidiametres, but the lowest poynt, is 166. semidiametres distãt from the earth.

Mercurie beeing lower then Venus, is variable and apte to bee changed, bright, but not white in colour; and is caried about the sunne like to Venus, as one whiles mouing before, and an other whiles following the sunne. Nor is hee further distant in the morning from him, then 29. degrees, and 37. minutes, and at the euening west-warde, 27. degrees, and 37. minutes. He perfourmeth his whole course, in the space of a yeare, as the sun doth. Also he is iudged to be the seauenth part of 21. or 22000. parte of the earth. Albategnius affirmeth, the Star of Mercurie, to be least of all the starres, and supposeth or accounteth him to be as one part, of 19000. parts of the earth. The highest ascention of Mercuries Sphere is from the earth after Albategnius) distant 166. semidiametres, but the lowest point in the same Orbe, is 56. semidiametres distant from the earth. The

The Moone being lowest of all the Planets, doth compasse about the whole Zodiack, in 27. dayes, 7. hours 43. minutes, and 7. seconds. She is lesser then the earth (after the iudgement of Ptholomie) by three hundreth nine times, and a vnity more then eight parts. For the triple proportion of the diametre of the earth vnto the moone, by deuiding aboue the sixt parts, is euen the like, as 27. vnto 5. But lesser she is then the sun, by sixe thousand, fiue hundreth, thirty and nine times. Copernicus (by his obseruations) founde the earth greater then the Moone, by forty three times: lesse then an eight part: and of this, the sunne also is founde greater then the moone, by seauen thousand parts, lacking threescore seconds And the greatest distance of the new and full moone from the earth: after the mind of Ptholomie, is 64. semidiametres, and 10. scruples: but after later obseruations, 65. semidiametres, and 30. scruples. And the lowest to the earth, is 55. semidiametres, and 8. minutes.

The Moone digressing from the Sun euery moneth, and taking or receiuing a newe light as it were, in that she (is changed, & taketh a new light of the sun) doeth after encrease by little and litle, conceiuing dayly a bigger forme and light, vntill shee come in right line against the sunne; at what time she shineth with full light: after returning againe vnto the sun, she wareth olde by losing of her light by little and little: and in the contrary maner cometh vnto the like formes of light, vntil she comming vnder the beames of the sun, bee quite out of sight. Also for that the moone hath a body, partly thin, partly thicke, solyde, and shadowed; therefore is she not equally filled round about with the beames of the sun, but that the same halfe of her Globe or body, which turned againe in heauen (that beholdeth the sun) is it which shineth, and the other halfe turned away from the suns light, is that which shineth not, but remaineth shadowed.

That

of the Sphere.

That Heauen hath a round fourme and
to be carried circularly.

Irſt, heauen is equally diſtant rouñd about from the earth, and of this is heauen perfect rounde, after the definition of the Sphere. Which reaſon is thus proued; that if heauen ſhoulde haue any other forme then perfect round, then of neceſſity muſt the ſtars change their diſtances from the earth, what place vppon earth they ſhoulde purchaſe, as ſomewhere more, and ſomewhere leſſe they ſhoulde bee diſtant; and the ſtanding of them changed, ſhould alſo alter their apparant bigneſſe, in that they ſhould appeare greater being ſeene neere hande, and leſſer, being ſeene far off. Yet neither of theſe happeneth, but that they cōtinually kæping aſunder, are drawn about by vnchangeable ſpaces, and holding a like bigneſſe and diſtance, to all places of the earth.

That the ſtars about the quarters of the Eaſt or Weſt, appeare ſometimes greater, is not cauſed by reaſon of the ſhorter diſtance, but for that their beams in the vapours, which often times conſiſt in the ayre betwæne the ſtarres and our eie, are then broken; which breaking of them, cauſeth the body of the ſtar ſæne, to appeare much greater in the eie, then in dæd it is.

The first Part

That heauen is drawne circularly, is thus knowne; in that wée alwaies sée all the Stars, from the East into the West, to be drawne vpward, and that the hemisphere in our sight, is carried continually in distant cycrcles equidistant, neuer changing the standing or distance, one from another, neither in bignesse, as far as the iudgement of the eie can descerne, neither any whit lessoned. For they being drawue from the neather place (as from the earth) are caried by little and little. And after they be thus come vnto the highest of their iourney (as vnto the noon-stæde) they decline again by little and little, till they be brought down vnto the west quarter, and there set and hidden, vnder the earth: and these places and times, both of the risings and settings doe they repeate in certaine order. Therefore by these it appeareth, that they are drawne and carried by round.

By the second it is euident, that the Starres, which be néere the Pole Articke, are neuer hidden out of our sight, but are continually and vnformally drawne round about the Pole as the Centre: in such sort, that the stars neare to it make the lesser compasses, and the stars further off, doe define greater compasses. So that the starres fastened to their proper Orbs (as aforewritten) are cyrcularly caried. By which two motions of the stars, as well tending vnto the West, as otherwise; it plainly appeareth that heauen is drawn about and caried round.

manifest demonstration appeareth of the former argument, by this figure here following.

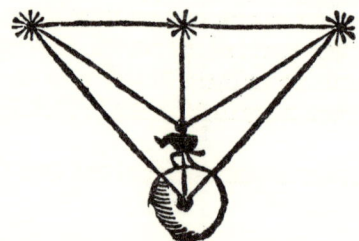

That the Water and Earth are round Bo-
dies, and by a mutuall embracing doe make one
Body, and one hollow
vpper face.

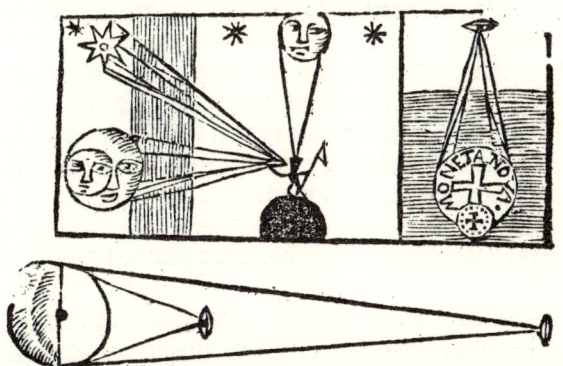

That the earth is round, is thus proued. Wheras in euery vpper face,
the length and breadth is considered. The length of the vpper face
of the earth, is from the West, into
the East, or contrariwise. The
breadth is from the South, into the
North, or contrariwise. That the
earth also to bee rounde, appeareth
after length: in that the Sun, Mone, and Stars, doe neither arise, nor set at one instant time alike, to all persons
dwelling in any parte of the earth. But doe much sooner
appeare and shine to them dwelling vnto the East, and
within a whiles after they shewe to them dwelling in the
West.

By the second appeareth, that one and the like Eclipse

of the moone in diuers houres, is séen both in the East and West. For that which appeareth in the first houre of the night to them in the West, is séene to them in the East parte, in the second, thirde, or fourth houre, euen as they come nearer vnto the East: which would not be caused, if the night to both places should happen and bée at one moment, nor sooner woulde they appeare to them in the East part.

Againe, there bée certaine stars, which in their rising, doe appeare sooner to them in the East parts, then to them in the West, as Plinie writeth of Arbelis (being a towne in Asiria) where an Eclipse of the Moone was séene in the second houre of the night, which in Sicily, was séene in the first houre of the night. For the Assirians are more Easterly then the Sicilians, and therefore doeth the sun set sooner with them, then with the Sicilians. And when it was also the second houre of the night in Assiria, the Sun first set in Sicilia, about the first houre of the night. Moreouer the Pole of the world (according to the diuersitie of places) is eleuated and depressed. So that the cause of the diuersity of this appearance, is onely the swelling of the earth.

To be briefe, the beginnings and spaces of the dayes and nights, and that in diuers places of the earth do vary, and yet following in a maner, one order. But this variety could not happen, if the earth were not Sphericall, and all about equally rounde, herein excluding both vallies, and the toppes of hilles, which applied (vnto the body of the earth) cause no inequalitie or diuersity at all. For the swelling of the earth causeth, that the stars be not séene togither in all countries, but drawne about by little and little, by a certaine succession and order, that they so appeare sooner to them in the East part, then to them in the West, through the swelling as yet not aboue caried, which swelling being high betwéene both, is a let and cause of the later appearing of them to the west: and by that meanes also

of the Sphere. 33

ſo keepeth and hideth the ſtars the longer from their ſight. So that by theſe it euidently appeareth, that the onely cauſe, is the ſwelling of the earth.

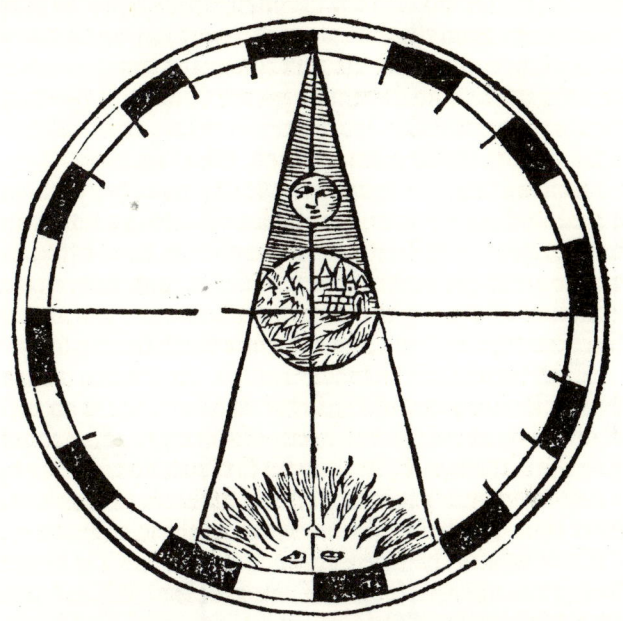

If the earth were faſhioned with a deepe hollowneſſe, and compaſſed round about with a light incloſure, then ſhould the ſtars riſen, be ſooneſt ſeene to them in the Weſt partes, and much later appeare to them in the Eaſt. For that the higher incloſure to the hollowneſſe, as a wal built about, ſhould be a let and hinderance to the ſight of the beholders; in ſuch ſort, that thoſe ſtarres ariſing, it ſhoulde hinder their ſight.

▸ If the earth were formed with places ſtanding in ſharp piller forme, or in right line vp; then ſhould the ſtars ap‐
D j.   peare

peare set, and be hidden alike to those places, and no differences of dayes should be caused, but that they shoulds haue one like day, and the sun also appearing to that side, which they shewed: so that whiles the Sun runneth and compasseth about the backe parts, they should be without light of the sun, and should remaine al the time in shadow and darknesse. And if it should haue a Cubicke forme, then should they see the sun sixe houres, and lose or be without light and sight of the sun, other eightéene houres.

If in round piller-wise, as if the bowndes were playne vnto both the Poles, and the hollow partes should decline vnto the East and West, then should no stars continually appeare to them dwelling in the hollow: but that certaine stars should arise vp and set in the West, and other certain stars néere to both the Poles, should alwaies be hid.

To conclude, if the whole earth were framed with an equal playnesse throughout, then should the stars appeare at one moment to all countries; and setting againe, should hide the like out of sight: and by that meanes shoulde the dayes begin and end alike, and no differences shoulde bée obserued. To all such arguments, seing experience onely doth repugne or contrary them: It is therefore manifest, that the earth from the West towarde the East, riseth vp into an equall swelling.

If the earth also were plaine from the East vnto the West, then shoulde the starres arise so sone to them in the West, as to those of the East, which is a manifest error. Also, if the earth wére playne, from the North vnto the South; and like from the South vnto the North: then the starres which were to some of a continuall appearance, should alwaies séeme the same and like, which way or into what quarter soever a man goeth, which also is vntrue. But the cause which maketh the earth séeme plaine, is through the ouer great quantity, which causeth it so to appeare to every mans sight.

But

of the Sphere. 35

But that the earth is round (according to latitude) the diuers eleuations of the Pole and stars (eyther alwaies in sight, or continually hidden) doth euidently declare.

For from the Equatour, in going forth easilie towardes the North, and that the Pole Articke be higher raysed, and the stars neere to the Pole raysed vp; then are the Stares right against like depressed, and as they were out of sight, and so much the more as they go further from the Equatoure: nor the Northerly stars neuer set, but continually drawne about (in sight) with heauen. But the contrarie happeneth, by going from the saide Cyrcle or Equatoure, vnto the contrary part. So that there is no greater cause of this diuersitie, than the swelling of the earth, which if the same shoulde bee plaine, the starres opposite or right against (according to latitude about the Poles) shoulde offer and appeare togither to all countries, which the swelling of the earth hindreth to be seene.

An inſtrument, by which the round neſſe of the Earth (according to latitude) may be proued, and all those may eaſily be ſhewed which are taught of the dayes *Artificiall*.

That

36 The first Part

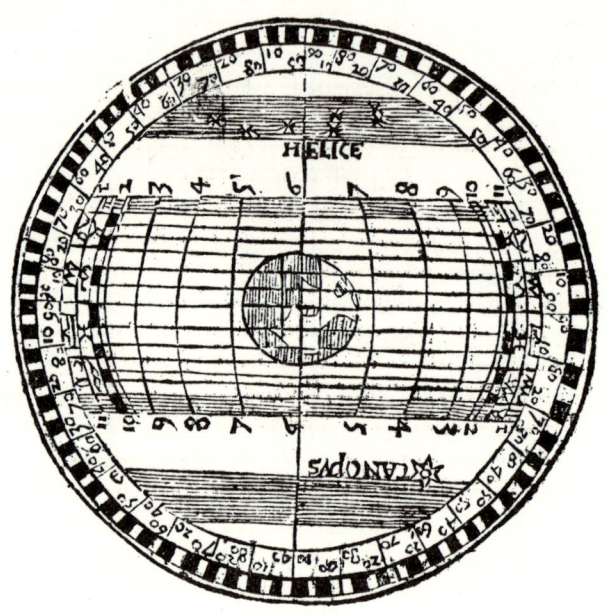

That the Water hath a like swelling,
and runneth round.

This by two reasons is proued, the first is most certaine, by a mark or marks standing on the sea banke, like as a tower, steeple, or such like erected of purpose, so that a shippe sayling into the deepe, and carried so far off, that no more of the sides or bottom can be descerned, sauing the top of the mast, which only appeareth

of the Sphere.

peareth to the sight. Or thus, that a marke stode on the sea banke, and a ship passing forth of the hauen, sayling so far into the sea, that the eie of the beholder being nære the fote of the mast cannot decerne the marke (the ship in the meane time staying or standing still) so that his eie being in the top of the mast, shall perfectly sée that marke: but the others eie being nære the fote of the mast, shoulde rather better sée the marke, than he which is in the top of the mast, as may more euidently appeare, by lynes drawne from either place vnto the mark: so that the manifest cause of this, appeareth to bee none other, then the swelling of the water. But here are all other impediments excluded that may otherwise hinder; as mists, foggs, and such like vapours ascending. Also a like reason of the impediments of this aboue written, is, for that the water ariseth into a swelling, which hindreth the sight of the bottom or sides of the ship (that being in a high place doeth not hinder the sight of the same) as the top of the mast, which either excéedeth or is equall with the swelling of the water. For men sayling on the mayne sea, sée nothing round about but the Sky and the Sea: but comming nearer the banks, do by litle and litle descry and sée, either high hilles or cliffes, as if they were rising forth of the water. Also to those that dwell on a high ground, the sun first ariseth, and last setteth. And to this agréeth, that out of the higher places, both more and further may bée séene into the sea, then in vallies or lower places. By all these therefore it is euident that the vpper face of the water swelleth, as by the example following more plainly shall appeare: but an other example of the same shall bée here rehearsed, by a similitude of one part as the whole. The similitude of which matter conceiue by this example, that experience dayly teacheth vs of the drops of water, which although they bée small, yet powred on drie wollen cloth, run into a round or bunching forme; which without doubt shoulde not be caused,

D iij.   if if the part folowed not the nature of the whole of his kind. Now the example aboue promised doeth here appeare, in which by the letter A. is the shippe ment to come vnto the marke C. In which being in the poynte A. that is in the bothom of the shippe, cannot sée the marke standing in C. through the swelling of the water. But he which is in the top of the mast, as in the poynt B. without all impediment may sée the sayde marke. That the selfe same or like to it, may be on land, as from the point D. none excepte hée bée foolish or starcke mad will affirme the like.

A

By the second it is manifest that the water by nature is caried and runneth downewarde, and slideth or falleth from higher vnto lower places, so long, vntill it hath filled and bee euen with the earth, through the staying of high heapes of earth, hilles or such like mighty and high banks inclosing it about that it run no further, nor make no hollownes in the middle of the earth, as a Center of the earth. Which therfore gathereth betwéene the empty places, so long, vntill it hath filled and be euen with the earth; and

that

that the whole togither, through the hollownesse thus made equall, doeth fashion and keepe a round forme.

So that the earth, with the sea, and waters running about it, do make one round body, and fill all the whole vpper face: the earth also gaping and open somewhere, receiueth water into those hollow places, but a parte of the earth appearing somewhere aboue it, staying and inclosing it about with strong inclosures and banks (wrought by diuine myracle) that the bare places of the earth, might be a commodious dwelling and feeding for all beastes, and other liuing creatures.

And y[e] this is true, shall bee prooued by other two reasons. The first, by sundry perygrinations, in which many and most large parts of the earth are found toward all the quarters of the worlde, which euidently witnesseth, that the earth is not as Plinie and others writeth; which imagined that the earth is compassed about with water, and appearing so out of the water, like an Aple or Ball swimming aboue the water, whose one halfe sheweth out of the water, and the other halfe hid in the water. Which reason Ptholomie doeth not allowe, but simply affirmeth, that the earth with the sea and waters, make one round body, by filling of the empty places, and both to haue one vpper face. Also Vitruuius in his ninth booke writeth, that the earth is placed in the middle of the world, and is naturally ioyned together with the sea in the place of the Center. But what the forme of the earth is, aboue the waters, is yet not throughly knowne, by reason of the sea which runneth betweene it in diuers partes, and breaketh it into sundry parts, like to gobbets or peeces.

Ptholomie affirmeth the earth to bee knowne, vnto the longitude of the halfe Cyrcle, that is 190. degrees, without any running betweene of the sea in that space; for that the earth is wholy ioyning together. But into latitude, he affirmeth the space to be much lesser, as 79. degrees, and of this opinion, is both Strabo and Aristotle.

By the second it appeareth, that the water with the earth, doeth equally make one hollow vpper face, and the same to be perfect round; but whether is bigger, is greatly to be doubted: although the learned Nouius, and sundry other late writers, doe affirme the face of the earth to bee bigger then the water.

By the third (which is the Ecclipses) In that of necessity the earth must haue such a forme with the waters running in it, as the shadowe of the earth frameth and counterfeiteth in the Moones Ecclipses; in that the shadowe sheweth and expresseth the forme of the bodie shadowed. But the shadow of the earth to be included round aboute with a round vpper face, the wise both sée & know. Therefore the whole Globe, compounded of the earth and waters, is comprehended with a round vpper face. For it is manifest, that the moone before and after the full, is séene horned, and the part shadowed of the whole cyrcle, is easily to be descerned from that bright circumference. So that the moone entring into shadow, or going out of the same, is likewise in the same maner horned: and the part darkned, is alwaies descerned from the cleare circumference of the whole Cyrcle imbossed. Therefore of necessity must the beginning of the shadow, which seperateth the parte lighted, from the shadowed, not bée fully straite, nor vnequal, nor vallied or winding, but round: and for that cause appeareth the vpper face of the shadowe not to bée plaine, but round.

By these is also manifest, as by the first, that there is no difference betwéene the Centre of the earth, and Centre of the water: but that the one, is the Centre of both the Elements, ioyning togither into one round body, and tending vnto the one and the same Centre of the earth. For the earth (séeing it is the heauier) is opened, and receiueth the waters falling into those places.

By

By the second it is euidente, that the place of the water, which ought to run ouer and couer the whole earth, is otherwise chaunged and appoynted by the Diuine will, for the benefit of all liuing creatures.

By the thirde appeareth, that the opinion of certaine Peripateticans is false; which affirmeth the water to bee ten times greater then the earth; and that to one parte of the earth, is ten portions of the water increased. But seauen times greater then the earth it cannot be, vnlesse the earth rounde about were wasted and impayred, by the Centre of the grauity (as it were setling and resting vpon) should yælde, and giue place to the waters, as the heauier. Seeing the Spheres are togither in a triple reason of their measures, then if the earth were an eight part, to seuen parts of the water, the diameter of it could not be the greater; as from the Centre of the waters vnto the circumference of them: that is, by double so much vnto the diameter of the water, as by this figure here vnder drawn appeareth: where this letter A. is the Center both of the earth and water, B. the Centre both of the magnitude and earth, G. A. D. the diametre of the waters, A. B. D. the diameter of the earth. If the waters are seuen times bigger then the earth, the diameter of them must nædes bæ double so much vnto the diameter of the earth, as hære from G. D. vnto A. D. By which example thus drawne, the whole earth receiueth the Center of the waight, gyuing place to the waters, and all couered with waters; to which generall experience gainsayth and denieth, muchlesse therefore can it be greater ten times. By which is to be concluded, that the water is but litle in quantity, in respect of the earth, although it may sæme very bigge, being vp to the edges of the vpper face of the earth. And if the waters had bæne more bigger then the earth, they had drowned or couered the whole earth, euen of late yeares.

That

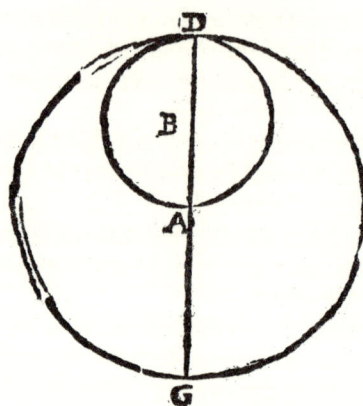

That the earth employeth the middle place of the Worlde, and is the Center of the whole.

Ristarchus Samius, **which was 261 yeares, before the byrth of Christ, toke the earth from the middle of the world, and placed it in a peculiar Orbe, included within Mar-ses and Venus Sphere, and to bee drawne aboute by peculiar motions, about the Sunne, which hee fayned to stande in the myddle of the worlde as vnmoueable, after the manner of the fixed stars. The like argument doth that learned** Copernicus, **apply vnto his demonstrations. But ouerpassing such reasons, least by the newnesse of the arguments they may offend or trouble young students in the Art: wee therefore (by true knowledge of the wise) doe attribute the middle seate of the world to the earth, and appoynte it the Center of**

of the Sphere. 43

of the whole, by which the risings, & settinges of the stars, the Equinoctials, the times of the increasing and decreasing of the dayes, the shadowes, and Ecclipses are declared.

The earth round about is equally distant from heauen: therefore, according to the definition of the Center, the earth is the Center of the world.

That the stars haue alwaies one bignes, in what place soeuer any shall beholde them: therefore are they in an equall distance from the earth.

The roundnesse of the earthly globe, hath a proportion vnto the roundnesse of heauen; that is, the certaine and proportionall parts in the earthly Globe, doe answere to certaine proportionall partes of heauen: therefore is the earth the Center of the world.

In that fifteene Germaine miles on earth, doe answere to a degree of the Meridian: and that in euery houre doe fifteene degrees arise of the Equinoctiall; which coulde not be, if the earth were not in the middle of the world. For the vnequall Arks, should otherwise appeare in the equal times: and the equall partes of the Meridian, shoulde the vnequall spaces on earth answere: which experience dayly witnesseth vnto the contrary. And hereof it ensueth, that the earth stands in the middle of the world.

In euery Artificiall day, doe sixe signes appeare, and sixe like set vnder the earth: therefore is the earth in the middle of the worlde, and is also as a pricke, to which the halfe doth regularly moue dayly. The like is in the opposition of the Sunne and Moone, when either light is in the Horizont: which could not be, if the earth should approch or come neerer vnto one part, then vnto the other.

If it were neerer to eyther of the Poles, then could not the vniuersall Equinoctials bee: for that the one Arke alwaies (either in the day and night time) should be greater then the other. The Ecclipses also coulde not bee in the
chan-

changes and full moones: For that there shoulde then bée vneuen spaces from the South vnto the North, and from the East vnto the West.

If the earth were not as the Center of the worlde, then of necessity shoulde these ensue, that the earth shoulde approch, either néerer to the East, or West, or South part: and when any of the starres (aswell the fixed as Planets) shall come vnto that part, they shall appeare nearer to vs, then being in any other part of heauen: and by that abouesaide, they shall also appeare greater: which is altogether vntrue, and we also sée the contrary in that (as aboue writen) they alwaies appeare of one greatnesse, eyther being in the East, or in the West. Also one halfe of heauen is alwaies aboue the earth, and the other halfe vnder the earth: and this is not onely found and knowne in one quarter of the earth, but the like in euery place (as the Equinoctials do witnes) then which there can be no more euident tryal.

A third reason may bée alleadged, if any imagined the earth vpon the Center, to be parted into two equal halfes, and that the eie is placed in the Center: then shall the eie sée no more then the halfe of heauen. By which appeareth, that the swelling of the earth, from the Center vnto his compasse about, in making a comparison vnto heauen is as in a maner nothing.

And it is knowne to the learned in Astronomic, that any of the fixed starres, is by many times greater then the earth: which if any behold them, they appeare as poynts in heauen. Now how much lesser would the earth appear, if a man should behold it from his place.

Here learne by this demonstration following, that the earth standing without the Center, in the poynt B. being to the Meridiane, as is the poynte A. nearer; and when a star shall come vnto that poynte, then shall it bée nearer to the earth, and in the opposite poynt, as is I. shall bee from the same much further, than in any other place, and shall

euen

euen there leſſer appeare, which by experience, is quite contrary. Further graunt that C. D. be the thwart Horizont, yet the contrary, for the ſecond reaſon, E. B. K. being the Equatour, which from the ſaid Horizont is deuided into two vnequall parts, and by this conſequent alſo muſt the Zodiacke bee deuided into two vnequall partes, from the ſaid Horizont: for that thoſe two Cyrcles, (as hereafter ſhall bee taught) doe croſſe one another into equall parts.

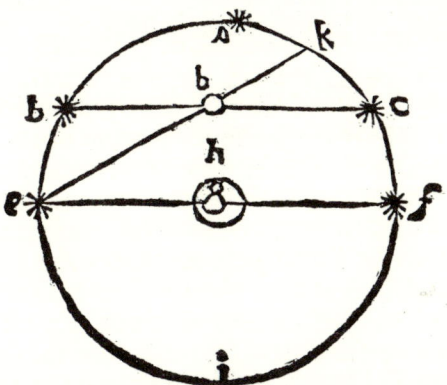

Therefore when the ſun, by his proper motion, carried from the Eaſt into the Weſt, ſhall come vnto the croſſings of the Equatour and Zodiacke, and that the greater part of theſe Cyrcles ſhall be vnder the earth, it cannot be that the Equinoctium or a like day and night, can bee throughout the earth, no not vnder the right Sphere, much leſſe can it be vnder the thwart Sphere.

If this be vnpoſſible, it ſhall be alſo as vnpoſſible, that ſixe ſignes may alwaies bee aboue the earth, and the other ſixe vnder the earth, but rather that more of the ſignes ſhall be vnder the earth, and more of them aboue: euen

as

46 *The first Part*

as the earth is imagined to be deuided from the Horizont aforesaid, into two equall parts; like as when it shall bée in the Center of the whole, and that from each Center of the greater Cyrcles, the earth is deuided into two partes. As all these (to any beholding the materiall Sphere) are forthwith knowne at the first sight: so by a third reason is to bee noted, that when any imagineth, by the lyne E. F. that the earth in the poynt G. standing as in the Center of the whole, is deuided by the middle, as well beeing in G. as H. for the excéeding distance from the Cyrcumference, is vnpossible to sée alwaies the halfe heauen.

If the Earth be not in the middle of the Worlde, then of necessitie shall it possesse some of these *standings.*

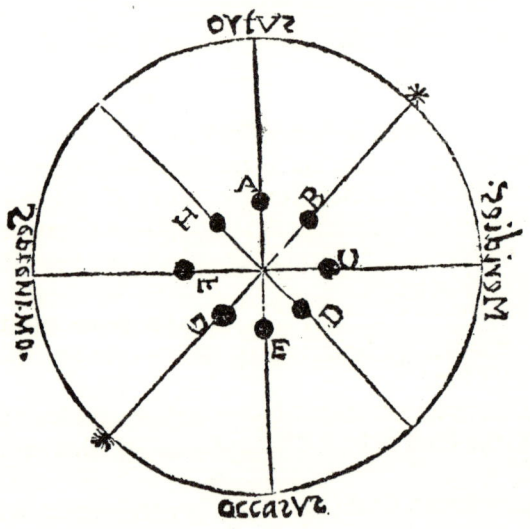

of the Sphere.

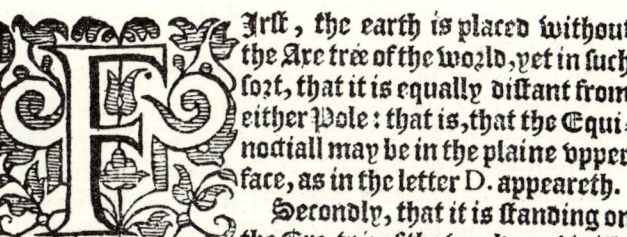Irst, the earth is placed without the Axe-tree of the world, yet in such sort, that it is equally distant from either Pole: that is, that the Equinoctiall may be in the plaine vpper face, as in the letter D. appeareth.

Secondly, that it is standing on the Axe-tree of the world, yet without the playne vpper face of the Equinoctiall; that is, that it be néerer to eyther of the Poles, as in the poynt B. or G.

Thirdly, that it is neyther standing in the Axe-tree of the world, nor in the plaine of the Equinoctiall, as in the poynts, A. C. F. E.

The first standing béeing graunted, these absurdities shoulde then ensue, through the diuers placing, in diuers and sundry places of the playne Equatoure.

1 In the right Sphere, should neuer the Equinoctium (or a like day and night) be caused, in that the Horizont shoulde neuer cut or part the Equinoctiall into two equall halfes.

2 In the thwart Sphere, shoulde no Equinoctiall bee, and somewhere againe should the Equinoctiall be, but not in the middle Parallell, betwéene the two Tropickes; that is, it should not happen the sun being in the Equinoctiall, but in an other lesser Parallell Cyrcle, being néerer to eyther Tropicke poynt.

3 The time from the rising, vnto the Noone tide, shuld not be equall to the time from the Noone tyde vnto the setting of the sun.

4 The magnitudes and spaces betwéen the fixed stars, both in the East and West, shoulde not be séene equall or a like.

In the second standing, if that the earth should bee placed

48   The first Part

ced on the Exe trée, and not in the myddle of the worlde, then should these absurdities ensue.

In euery Climate, the playne of the Horizont shoulde cut heauen into two vnequall halfes, except those places hauing the right Sphere: yea, and the Zodiacke shoulde be deuided into two vnequall Arks, so that there should be somewhere more, and somewhere lesse then sixe signes of the Zodiacke séene aboue the Horizont, which is contrary to all experience.

2 The Equinoctiall shadowes, both of the rising and setting of the sun, shuld not agrée, in such sort that they might fall in right line. Neither the shadowe of the rising of the sun in the Solstitial or longest day, should make or stretch in right lyne, with the shadowe of the suns setting, in the Brumall or shortest day, *et econuerso*.

In the third standing, if neyther it should be on the Exetrée of the worlde standing, nor in the plaine of the Equatour; then should the same absurdities ensue, which are vttered in the two former.

To conclude, wheresoeuer the earth is generally placed without the myddle of the world, there is the reason of the dayes equal increasing & decreasing in the thwart Sphere confounded, and there shall eyther no Equinoctials at all

be

bee caused, when the sunne occupieth the myddle way betwéene either Tropicke. Nor the Moone alwaies shadow the suns light, although she commeth right against the bodie of the sun. And the earth not standing in the myddle of the world, shall not shed or stretch his shadow to the moon. So that all these absurdities and vaine argumentes doe grant, that the earth cannot bee in any other place, then standing in the middle of the world.

That the Earth abideth fixed and vnmoueable, in the myddle of the world.

That neyther the earth, in right nor Cyrculare motion is drawn about the Exe trée of the world, nor about any other Exe-trée, but to rest and stay in the myddle of the worlde; both holy scriptures confirme, and Phisicke reasons prooue. For the Psalme sayth, which stablished the earth vpon his foundation, that it shall neuer bee moued. And Ecclesiastes in the first chapter sayth: that the earth standeth for euer, and the sun both riseth, setteth, and goeth about vnto the place where he arose. Also that the sun is drawne about, the Psalme doth manifestly witnesse, where it is saide: that for the sun, hée hath placed a Tabernacle in them, and he, as a Bridegroome going forth of his chamber, doth reioyce as a Gyant to runne his course, which goeth forth from the vttermost bound of heauen, and returneth about vnto the ende of it againe. Also it is knowne and numbred among myracles; that God would haue the sun to continue.

The Phisicke reasons are these.

That of one simple body, is onely one simple motion. That the earth is a simple bodie: therefore thereunto agreeth but one simple motion.

But of the simple motions, I haue before taught, that the one is in right maner, and the other in Cyrculare forme. That the right motion seeketh downwards vnto the myddle, whether being caried, they settle and rest. Therefore is the motion of the earth not cyrculare about.

By the second appeareth, that euerie graue or heauie matter by nature, is through his waight caried after a most straite lyne vnto the Center, and both setleth, stayeth, and resteth at the same; where it neither falleth, or is caried any further. So that all graue matters, as the parts of the earth, and those which consist of the earth, are sent or caried by a most straight leading vnto the earth, and at his vpper face shall stay and rest. And weare it not that they are staied through the fastnesse of the earth, they should so long be caried downwardes, vntill they came vnto the Center. Also the earth through his fastnesse, receyueth and beareth all thinges falling on it. Therefore doeth the earth much more (beeing within the Center) stay and rest fixed and vnmoueable, bearing all other heauie things falling on it, seeing the earth is heauiest of all others.

By the third it is euident, that if the earth shoulde bee moued or caried, it should of necessity be either drawne in right, or cyrculare motion. If it should be caried in a right maner (seeing it is the heauiest of all others) it shoulde by his swiftnes moue and goe before all other heauie things, and shoulde leaue behinde the liuing creatures, and other thinges fastned to it, and shoulde also leaue them hanging
be-

of the Sphere.

behinde in the Ayre.

If the earth should be drawne about by a cyrculare motion, it should in a daies turne (at the least) be caried about the Exe-tree, from the West into the East, as either alone or with the first Orbe: then euery day, should many most disordred things, and contrary to experience happen. For it shoulde bée a most spéedy motion, and swiftnesse inseperable, which should draw cyrcularly all the whole earthly body rounde about in 24. houres. And therefore that the earth is caried with so swifte motion, shoulde not onely ouerthrowe buildinges, but high hilles, and greatly shake and harme all thinges fastned and growing on the earth: yea all liuing beasts, and other creatures dwelling on the face of the earth, shoulde bee likewise shaken and harmed. Also the cloudes, foules, and whatsoeuer liueth and hangeth in the Ayre, should bée caried and lefte behinde at the setting in the West. For by the swift turning about of the earth should all things be ouer turned, and left behind by a greate and long space; if by such a swiftnesse, the earth should be turned about the Exe-tree of the world. Or if by the motion of the earth, the ayre, and all things hanging in the ayre, should be drawne with a like swiftnesse; then should they appeare to stay, or not to be moued at all. And further, if a stone or any waighty thing cast vpward shuld not light againe downe right on the same place: as may be séene in a shippe, at sayling. So that to all these, doeth euident experience deny, that by no motion, the earth is any thing moued, but continually stayeth and abideth.

By the fourth it is manifest, that in the motion and turning about of the Cyrcle, the Center abidsth vnmoueable: which is the earth, placed in the myddle of the world, and is as the Center of the worlde. Therefore is the Earth knowne to be vnmoueable.

℃ ij.   That

That the Earth compared vnto Heauen, is as a poynt.

Lthough to the vnskillfull in this Arte, the magnitude and largenes of the earth seemeth to be of an excæding greatnesse, that no bond or ende can be decerned with the eie; nor any hauing trauailed into farre countries, could hytherto finde any bounds of the same: yet the greatnesse of the earth compared vnto the mighty largenesse of heauen is accompted but a pricke; as the Geomitricall rules declare.

The earth also is a very small thing, in respect of heauen: yea so litle in comparison, as a pepper corne, or sæde of Colliander, vnto a Cyrcle of a thowsande paces compasse.

For if the earth compared to the firmament, were of any sensible greatnesse, a man shoulde not sée the halfe of heauen, nor the halfe Cyrcle of the Equinoctiall or Zodiacke. And howe much greater the earth shoulde be, by so much the lesser should a man sée the halfe of heauen. But the contrary is knowne, in that on any plaine of the earth or vpper face of the sea, a man alwaies séeth the halfe Sphere of heauen, the other halfe (in the meane time) remayning hid: and of this the halfe dyameter of the earth is so small, vnto the distance of the firmament, that it may take away nothing in a maner of the halfe Sphere, extant to the eye.

Besides these, if the earth should be imagined to be placed in any of the Orbes of heauen, it woulde appeare but small in respect of them: for being imagined in the Moones Orb; the earth should appeare thrice as great as the moon is decerned from thence, and somewhat bigger. And from the suns Orbe, the earth should be decerned twice so large as

of the Sphere.

as Venus doth here appeare to vs. And if in Marses Sphere you would say that the earth is equall to a small star: But from the firmament, Saturns Sphere, or Iupiters (if a man could decerne it) the earth shoulde appeare so small, that a man would be abashed at the sight of it.

And here an ignorant man, might greatly wonder, that so small a body, yea rather a pricke (as it is accounted of the learned) should containe in it so many Realmes, Prouinces, Citties, Towns, Flouds, Mountaines, Woods, Valleyes, Seas, Riuers, Lakes, and many other greate matters, ouer long to be written.

That the earth also is as a Pricke, is declared by sundry reasons following.

By the first, that round about the earth, the magnituds and distances of the stars, in their times, are decerned and seene euery where equall and alike.

By the second, that the Gnomons or dyall shadowes, and the Centers of the Sphericall borderes or Cyrcles, placed in any part of the earth, do somuch auaile, and kéep the considerations and guydings about of shadowes, so regularly, and agréeing to the rule and matter, as if those (in very déede) shoulde bee placed in the myddle poynte of the earth.

By the third, that the Horizont doeth euery where deuide the whole heauen into two equall halues. For that in euery moment, doe sixe signes of the Zodiacke appeare aboue the Horizont, and in the night time being a fayre sky are they to bée séene with the eye, and so many (at that instant) hid within the Horizont: so that by a cōtinuall drawing about of heauen doe sixe signes appeare, and as many right against those sire, set vnder the earth. If the magnitude of the earth shoulde bée of any light portion vnto heauen, then so much of the Center (the vpper face drawne aboute) shoulde parte or deuide heauen into equall halfe Spheres.

E iij.　　　　　　The

The first Part

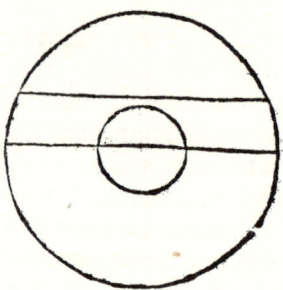

The other Spheres reatching from any part of the vpper face, shoulde deuide the same into vnequall portions. Neither halfe the Zodiacke should alwaies appeare, but a portion, much lesser then halfe the Zodiack should be seen aboue the earth, so that the greater parte of the earth, through the solydnesse excluded and hidden, should not after be seene.

By the fourth, the Equinoctiall shadowes, both of the rising and setting of the sun, doe make a right line, euen as if they should be streached out and lie on the plaine, caried by the Center of the earth. So that all these should not be caused, if the magnitude of the earth in respect of heauen, should be of a sensible, or of any portion to it. To conclude, Ptholomie vseth alwaies the body of the earth for the Center of the worlde, not deuiding the vpper face from that which is not in sight of the earth.

Certaine affirme, that one degrée of the greatest Cyrcle in heauen contayneth 57051. common Germaine miles. Of which one degrée of any earthly Cyrcle in the vpper face of the earth, doeth amount to 15. Germayne myles. And that one minute of the celestiall degrée, expresseth, 9509. Germayne miles, which (if this bée true) and certainly knowne, then is it not vainely thought and gessed, that the earth is as a Pricke, in respect of heauen.

To

To finde the compaſſe of the Earth, and by it the Dyameter.

The whole compaſſe of the earth, according to Ambroſius, Theodoſius, Macrobius, and Eratoſtenes, doth contayne 5400. Germaine myles, and the dyameter of the ſame, doth contayne 1718 $\frac{4}{22}$. Germayne miles.

But the authority of Eratoſtenes (after the minde of Plinie) is more to bee regarded, then the other three Philoſophers, which prooueth by demonſtration and reaſons, that the compaſſe of the whole earth is, 252000. furlongs.

Yet Hipparchus finding faulte at Eratoſtenes, doeth affirme the compaſſe of the earth, to be of 277000. furlongs. And a furlong is here (after the agrément of the Geometricians) of a hundreth, twenty, and fiue paces.

And this ſentence is not here mente, that there is any ambiguity or vncertainty in this reaſon: but that the one affirmed leſſer, and the other more furlongs. For after Eratoſtenes, doe 700. furlongs anſwere to one degrée: but after Hipparchus, 774. furlongs anſwere to a degrée. So that there is no other diuerſity in the matter, but onely the number.

Ptholomie that was after Eratoſtenes, attributed ſeuen hundred fifty furlongs of the earthly meridiane, to one degrée of the celeſtial meridiane. So that by all theſe appeareth, that the magnitude of the earth is as yet vnfounde out, through the difficultie of meaſuring. And this whole compaſſe is not onely ment of the earth, but of the earth and water ioyntly togither, both which are ſaide to make one Sphere.

Also Eratostenes gathereth the compasse of all the earthly Orbe, by the proportion of the perticular, or the degrée of the celestiall Cyrcle, vnto the like space on earth. For he affirmeth, that to one degrée of the celestiall Equatour, answere 700. furlongs, or 15. Germayne myles, but Ptolomie attributeth to a degrée, 500. furlongs.

Which is thus to be vnderstoode, that a Cyrcle be imagined on earth, directly vnder the Equinoctiall or Meridian lyne, deuiding the earth into twoe halfes: and that this Cyrcle be likewise deuided into 360. parts or degrées as the celestiall Cyrcles are.

And ech of these parts doth like vnto the celestial parts, containe 700. furlonges, or 15. Germaine myles. This nowe being tryed and found, what the whole Summe eyther of the furlongs, or myles of the whole cyrcumference of the earth, which contayneth 360. parts or degrées: you shall easily finde and knowe the same by this maner. Multiply the whole compasse of the earth; that is, the 368. degrées, by the 700. furlongs, or fiftéene Germayne myles, and the whole compasse shal either appeare to be 252000. furlongs, or 5400. Germayne myles.

This whole compasse of the earth, deuide by 22. and the number comming thereof, shall bee the 22. part of the compasse of it; that is, $11454\frac{12}{22}$. furlongs, or $254\frac{12}{22}$. Germayne myles.

And abate this 22. part, from the whole Summe of the circumference, and the number in furlongs shall remaine and be $240545\frac{10}{22}$. and in Germayne miles $5154\frac{22}{22}$.

And if any of these sums be deuided a part by 3. it shal be found in furlongs to be 80181. a halfe, and a third part, or $3.\frac{10}{66}$. And in Germaine myles $1718\frac{4}{7}$. that is; the dyameter of the earth, aswell in the furlonges, as Germayne miles.

And Archimedes by sundry labours, and witty inuentions, and by Geometrical practise, hath found, that the like

of the Sphere. 57

proportion is of the Circumference of the whole Cyrcle vnto the diameter of the same, as is 22. vnto 7. that is, the diameter thrice, with a seauenth part and a halfe.

But whensoeuer any man will (by the cyrcumference of the Cyrcle) gather and finde his diameter, worke the numbers thus, as this example teacheth. First, set down 22. at the left hand, toward the right hand 7. and the cyrcumference between those two numbers, 22. 5400. 7. After multiply the first by the second, that is, 7. by 5400. the number increased, which is, 47800. deuide by the thirde, that is, 22. and you shall finde in the quotient, 1718.4/22. Germayne myles.

Or thus in furlongs, the number being set downe alike 22. 252000. 7. then multiplie the first by the second, as 7. by 25200. and the increase shall be 1764000. after the increased number: deuide by the third, as by 22. and the diameter shall be, 80181 18/22.

If any couet to finde the vpper face of the earth, by the dyameter, and cyrcumference known, worke one into the other, and you shal haue that you seeke. But if you desire to knowe the thicknesse of the earth, then ioyne the superficiall solydenes of the Sphere, vnto the sirt part of the diameter, and you shall obtaine your desire.

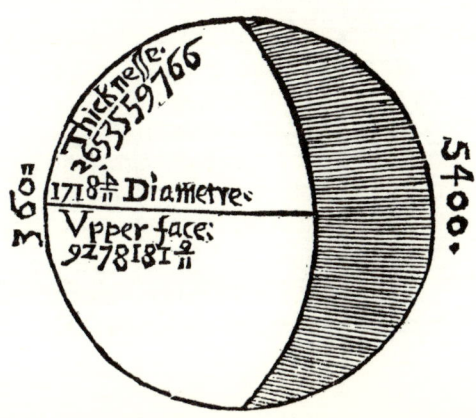

THE
SECOND PART
OF THE SPHERICALL
Elements of the Celestiall Circles, with the vses of the
same Circles.

What is the Summe of this
Second Part.

Whereas in the first part, were only the rudiments of the Sphere handeled and taught (which are also written and contained in diuers Physike bookes) as of the World, and the many parts thereof: that is, of the Ethereall and Elementarie Region: And also of the parts, motion and forme, of the Etheriall Region: as Heauen, and the forme, standing, and quantitie of the Earth.

Here in this second parte, shall fully bée set foorth, and largely handled, the manifold vses of the Cyrcle, of which
the

the materiall Sphere is framed and made. Further this second part is deuided into thrée partes, the first teacheth the deuision of the Cyrcles (in that the aunciente Astronomers, for a playner instruction, deuided heauen into sundry Cyrcles) and of these some in greater, and other some in lesser Cyrcles. In the second part, are the definitions, descriptions, and vtilities of all the Cyrcles taught. In the third and last part, are the places of the Zones, learnedly described, and the vtilities of them.

So that this second part doeth especially intreate of the Cyrcles (séeing the principall poynte of the Sphere, is of the celestiall appearances) which by reason of the celestiall Cyrcles, or of the first moouer are caused; as may appeare of the ascentions and descentions of the signes, by which the whole knowledge aswell of the naturall as artificiall day is learned.

Wherefore in that this instruction of the ascentions of the signes, consisteth in the Cyrcles (which the auncient Astronomers imagined to bée in the first mouer) therefore is this second part of the celestiall Cyrcles, aptely placed, and necessarily before taught.

That the Sphere of the worlde, is either right, or thwart.

The roundnesse of the earth, as is afore taught, both altereth the standing of the Poles, and the whole Sphere of the worlde, in diuers partes of the earth. For to them which dwell vnder the Equatour, either Pole falleth to the playnesse of the Horizõt. But to others dwelling without the Equatoure, the
one

of the Circles. 61

ne Pole is raysed, and the other depressed & hid: through which diuersitie of the standinges of them, are these differences caused; that the risings and settings of the signes are altered; the spaces betwéene the dayes and nights varied, whose causes ought diligently to be sought. Therefore is the right Sphere, distinguished from the thwart Sphere of the worlde. In this maner, as here you may beholde by these figures following.

That is called the right Sphere, in which either Pole resteth and standeth on the plaine of the Horizont and the Equatoure, which there doeth exactly possesse the middle place betwéene the Poles, and doeth with the Horizont make a right sphericall angle: of which it is so named a right Sphere. For they haue such a standing vpon the Sphere of the worlde, as that neyther of the Poles is eleuated aboue the Horizont, to them which dwell vnder the Equatoure.

And to cut at right Angles, is none other, than so to cut a Cyrcle into a Cyrcle, that the Angles which are caused on eyther side, are alike equall, as this parte of the Equinoctiall, A. B. C. and this part of the horizont D. A. E. which is crossed of the Equatoure in the poynt A. and the Angle B. A. D. is equall to

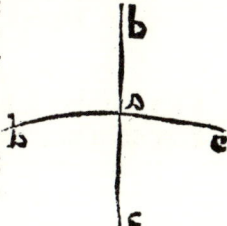

the Angle B. A. E. therefore shall B: A. crosse into D. A. E. sphericall wise perpendicularly, and the Angles D. A. B. and E. D. B. shall bee both right, by the definition of the right angle, as was before taught.

The thwart, declined, or bending Sphere, is that, in which either of the Poles of the world eleuated, is seene aboue the Horizont, and the other iust somuch set and hidde beneath the Horizont: and also that the Equatoure frameth and maketh with the Horizont thwart and vnequall angles. And that is called a blunte angle, which seeth the Pole eleuated: and that a sharpe angle, declining vnto the contrary.

They which dwell on this side, and beyonde the Equatoure haue such a Sphere. But the same forme and condicion of the thwart Sphere, is not euery where; nor the positure of it, the same reason: but that the thwartnesse of the Sphere is so much the more increased, as by how many degrees either of the Poles are neere to the earth: and beeing further distant from the Equatoure, is raysed and caried higher, which is the cause of many obscure differences: which that they may the plainer be expressed and vnderstanded, the skillfull practisioners haue deuided Cyrcles in the first mouer, by lynes drawn vnto certaine stars or prickes from the Center of the earth, and drawn about either by a continuall or dayly motion, by which they imagined them to be described.

That the Circles of the Sphere, be some greater, some lesser, and the number of the Circles.

HEre it is not to be omitted, that one Cyrcle is greater then another, by foure meanes. First, by reason of the magnitude of the celestiall body in which it is imagi-

of the Circles.

imagined to be. And of this is the Equinoctiall Cyrcle of the first mouer, greater then the Equinoctial Cyrcle of the eight spher, in that the first mouer is greatest of all the bodies. And although the Equinoctiall of the eight Sphere, doth deuide it into two equall halfes, yet of the first mouer it is named the greater, for that the same includeth all other bodies.

By the second, it is euident that the Equinoctiall Cyrcle is greater, by reason of the appearaunce, in that the whole is seene aboue the Horizont. And by the same reason the Northerly Cyrcle (which is named the Arcticke Cyrcle) is the greater, for that it alwaies appeareth to vs, aboue the Horizont.

By the thirde, the Equinoctiall is accompted greater then the other, in regarde of the influxiue vertue: and for this cause also is the Zodiacke called greater then the others, through his greater working into these inferior bodies. For that vnder it, the sun and all other Planets are drawne. And Hipparchus writeth, that this Cyrcle is the life of all thinges which are in the world, &c. In that by the ascending of the sun to vs, generation is caused, and by his falling or going from vs, diminishing, that is corruption getteth the vpper hand.

By the fourth, is a Cyrcle called greater then the other, insomuch as it is one Sphere, and thus the equinoctiall, is greatest of all the Parallell Cyrcles, in the first moouer: which is euidently demonstrated, by the diameter of the Cyrcle. Therefore by the definitions and reasons aboue shewed, the equinoctial is the greater Cyrcle, described in the vpper face of the first mouer, according to each part, or the whole of it, beeing equally distant from either Pole of the worlde.

And it is further to bee considered, that all the Cyrcles of the Materiall Sphere, are imagined to bee in the first mouer, which also a materiall Sphere doeth especially represent

present. So that these Cyrcles, may also bee imagined in the other Spheres, aswell as in the eight Sphere, &c.

And although a man may enter into conference betwéene these Cyrcles and the diameter, yet he shall be forced to confesse that they be on such wise vnto the sphere, as the Cyrcle is vnto the diameter. So that as the diameter deuideth the Cyrcle into two equall partes (in that it passeth by the Center of the same) euen so doeth euery of the greatest Cyrcles deuide the Sphere into two equal parts, because the playne vpper face of it passeth by the Center. And by this it may easily bee perceyued, that those which are named the lesser Cyrcles (of which is a farre greater number than is here set down) haue diuers Centers from the Center of the Sphere; and yet the playne vpper face of them passeth not by the Center of the same Sphere. Of which ensueth, that they cannot deuide the sphere intotwo equall halfes: no more then the lyne drawne without the Center, into a Cyrcle; can deuide the same into two equal halfes. And both the greater and lesser of these is mente, according to the distance of his Center, from the Center of the sphere.

The inward Cyrcles that be mouable, are those, which are descrybed in the first moouer, and are drawne with it about: as is the equinoctiall, the Zodiacke, the Colures, the Tropickes, the Polare Cyrcles, and others descrybed from the poyntes of the first moouer. But the outwarde Cyrcles, are they that are as immoueable, and not drawn about with the first mouer, but abide steady. The number of which are these: the Meridiane, the Horizont, the houre Cyrcles, the verticiall Cyrcles, and Cyrcles of the progressions.

Further it is to bee noted, that many are the Celestiall Cyrcles (as is aboue declared) whose vse partely vnto Astronomy, and partly vnto Astrologie, is necessary. As the verticiall Cyrcles, the Cyrcles of the altitudes, the Cyrcles

of the Circles.

kes of the celestiall houses. The Cyrcles with the which the materiall sphere is descrybed: and to bee briefe, there are so many celestiall Cyrcles, as there may bée poynts ymagined in the first mouer.

Yet are there but onely ten Cyrcles, which are required vnto this sphericall treatise; whose names are the Equinoctiall, the Zodiacke, the two Colures, the Meridian, the Horizont, the twoe Tropickes, and the twoe Polare Cyrcles.

The greater Cyrcles are those, which haue the same or a like Center with the earth, whose playn vpper face doth passe by the Center of the earth, so that they deuide the sphere, into two equall parts (and especially the equinoctiall) which for that it is a greater Cyrcle, doth cut the spher into two equall halfes; so that his playne vpper face passeth by the Center of the earth, according to the definition of the greater Cyrcles. And by this consequent, when the Sun is in the equinoctiall, he falleth into the Center of the earth; that is, hée is in the vpper face which passeth by the Center of the earth. And the sun is neuer in such an vpper face, but when he is in ÿ two equinoctial poynts for otherwise, he runneth without that vpper face. For the greater Cyrcles are a like vnto the Sphere, as the diameters vnto the Cyrcle: in that as the diameter cutteth the Cyrcle in two equall halfes (for that it doeth passe by the Center of the same) euen so doth the greater Cyrcle deuide the Sphere into twoe equall halfes, in that the playne vpper face of the same, doth passe by the Center of the sphere.

But the lesser Cyrcles are those, which haue diuerse Centers, from the Center of the sphere, so that the playne vpper face of them doeth not passe by the Center of the sphere. For how much nearer the Center of the same is to the Center of the sphere, and somuch the greater is that Cyrcle, as the Tropicke. But the further it is from the Center, euen so much the lesser in sight is the Cyrcle, as

F j.   are

are the Polare circle.

And here none may suppose, that either these or other like cyrcles, to be verily in the first mouer, but only to be vnderstoode or imagined. For the cause of deuiding heauen into certaine spaces and regions, through the helpe of which, the courses of the Planets are obserued, & brought vnto a rule.

Further the office of the celestiall cyrcles, are these.

1 That they deuide heauen, into certaine spaces, or regions.

2 The courses of the Planetes, the firmament, and first mouer, by the helpe of these cyrcles are obserued, and brought vnto a rule.

3 They shew the points of rising and setting; the nearnesse and differences both of dayes and nights.

4 The times and varieties of all the celestiall appearances, may bee obserued and knowne of the cyrcles by a certaine reason.

The sixe greater cyrcles are numbred by names, standing, and vse distinguished.

As the {
Equinoctiall.
Zodiacke.
Colure of the Equinoctials.
Colure of the Solstices.
Meridiane.
Horizont.
}

But there are many others, as the Cyrcles defined or descrybed by the Poles of the Zodiacke, and Centers of the stars, which are named the cyrcles of the latitudes.

The

of the Circles.   67

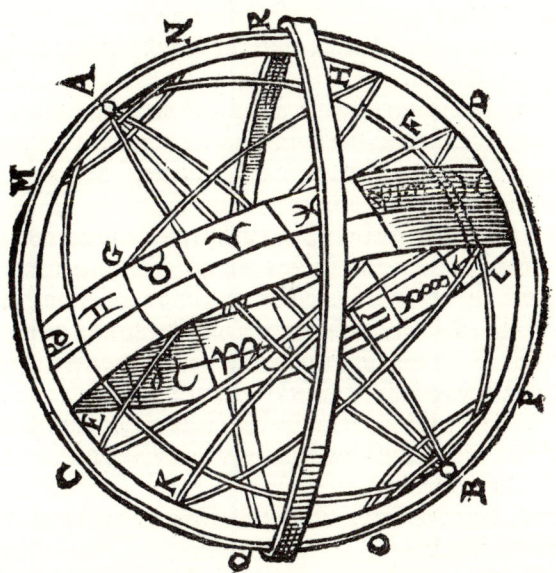

    The Cyrcles drawne by the verticiall poynts of diuers places, which may bee named the cyrcles of the diſtance, or ſpace betwéene places. For that the toppes doe knit or ioyne together by the nigheſt ſpace of the differences of places, and doe ſhew the diſtance of them.

    The cyrcles by the Centers of the ſtars, and Poles of the worlde drawne, are named the cyrcles of the declinations of the ſtars.

    The ſire cyrcles of the poſitions (through which by the thirty parts of the Equatoure, and the poynts, touched of the Horizont and meridian drawne ouer the Equatoure) doth Regiomontane part and deuide the whole heauen into twelue equall ſpaces, which hee nameth the houſes of heauen.

    The ſire greater cyrcles (through which by the Poles of the Zodiacke, and the thirty partes of the ſame bended

                F ij.   and

and wrythed) doeth Iulius Firmicus deuide the Zodiacke into twelue equall parts, but the Equatoure into as many vnequall arks.

But that former diſtribution and dispoſing of the Cyrcles by Regiomontane, both deuiſed and demonſtrated of him, doth bring and cauſe a reaſon of the framing of the figures of heauen, which they name Rationall: in that the ſame inuented and taught by principles and demonſtrations, is declared by certaine reaſons. The other inuented and exerciſed of others, doth bring and cauſe another reaſon of the forming and erecting of the figures of heauen, which of the ſame they name the equall maner; in that it parteth or deuideth the Zodiacke into equall arks. Many other cycles there are, which for breuity be here ouerpaſſed: for that they belong not vnto this determined treatiſe of the principles.

The leſſer cycles, although there are in a maner infinite of them, yet are there foure only recited and eſpecially knowne, which alſo are named Parallelles.

As the
{
Tropicke of Cancer.
Tropicke of Capricorne.
The Articke or Northerly Circle.
The Antarticke.
}

Of the foure greater cycles, afore written, they both are moueable, and are continually drawne about with the firſt mouer and neuer changed. But the two neather cycles, as the Meridiane and horizont, doe remaine and abide fixed and immoueable, in the going about of heauen, and the ſtanding alwaies changed on the earth, towarde what quarter ſoeuer they are varied, as they may be and are in a maner infinite in number.

The Aſtronomers deuide, both the greater and leſſer cycles, into 360. degrees, which they ſo named through
the

of the Sphere.   69

the suns passage or iourney in the Zodiacke, measuring and defining by his dayly course such like partes and spaces. And of these partes or degrées of the greater cyrcles, it is found and known, that each degrée contayneth in the vpper face of the earth, either 62500. paces, or 500. furlongs, or 15. Germayne miles. But each parte of the lesser cyrcles, doe comprehend a lesser space by somuch, as by howe much more from the magnitude of the Parallell, which is the middle and greatest, by reason of the distance, they lacke or differ.

And each of the thrée hundreth thrée score degrées, are also parted or deuided into thrée score minutes: and each minute, into thréescore seconds: and each second into thrée score thirds: and so forth from thirds vnto fourthes: and so vnto tenths, they distribute them.

The description, names, and vtilities, of the Equinoctiall.

He Equinoctial, which the Grǽks name ιϲημεϱιν, is a greater cyrcle, placed in the myddle place of the Sphere, betwéene either Pole of the worlde, and deuiding both by equall spaces moueable, and crossing the Zodiacke in two poyntes, which when the sun doth come vnto, hée then causeth a like day and night throughout the earth: whereof this cyrcle first purchased that name; in that the day is equall to the night, which happeneth twice in the yeare, as in the beginning and entraunce of the sun, both into Aries and Libra.

And a straight lyne drawne out by imagination doeth describe this cyrcle, reaching out of the Center of the earth by the Center of the suns body on a plaine or flat, of the E-

F iij.   qui

quinoctiall being vnto the first moouer, or vnto any of the fixed stars to the Equatoure, fastned to the eight Sphere, as that either of them, which with the thirde being somewhat lower and darcker doe fashion the gyrdle of Orione, and that by a dayly and continuall turning drawne about of the first moouer, vntill the same bee returned vnto the place, from whence it began.

And likewise the sun, or any other constellation placed or being in it, doth in euery region describe the halfe of the Parallell aboue the Horizont, and the other halfe, vnder the Horizont: which Ptholomie nameth the cyrcle of the Equatoure of the day: and Alphraganus, the cyrcle of the Equinoctiall, and swathe or gyrdle of the first moouer, in that it compasseth about the first moouer (as Strabo writeth) that it parteth the Northerly halfe Sphere, from the Southerly. For this greater cyrcle of the first moouer, is the first measurer, both of time, and motion.

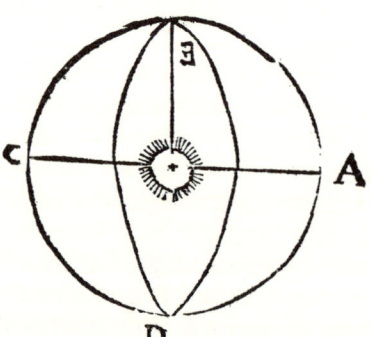

In that it causeth the proportionall cyrcumferences by the spaces of times: and whiles it is once drawne about, a naturall day is performed. And whiles the compasse also of the whole, haue meued one foure and twenteth part, an equall or Equinoctiall houre hath passed: by which it

doth

of the Circles.

doth euidently appeare, that this cyrcle, belongeth vnto the first mouer.

This Worthy Circle hath diuers names.

1. It is named the Equinoctiall, in that it causeth a like night, to the artificiall day.

2. It is by the same reason named the Equatoure; for that it maketh equall the night, to the day.

3. It is named the gyrdle of the first mouer (not vnproperly) for that as a gyrdle doeth gyrde or deuide our body into two equall halues; euen so this cyrcle deuideth the sphere or first mouer, by the middle.

4. It is named the line of the equality of the day, or the line of the equation of the Orbe of the day, or the iust denition of the day and night.

5. Of Plinie it is named the Center of the earth, and that not incongruently; seeing all the Parallell cyrcles described from the Center of the sun by the motion of the first mouer, haue their Centers from the Center of the earth: and that the Equatour onely, which when the sun shall be in the Equinoctiall poynt, is then imagined to be drawne aboute with the motion of the first moouer, that hath the same Center with the earth at that time, by which the playne of the Equatoure, is then noted to passe. So that this is the cause why Plinie giueth that name to it: seeing a like day and night is caused, the sun then running vnder the Equatoure throughout the earth, as no man of skill maketh doubt of.

It is named the cyrcle of the high solstice; but this commeth to passe, by reason of those which dwell vnder the equinoctiall

F iiij.

quinoctiall, and haue foure solstices; as two on hie, & two below, hauing foure shadowes in the yeare: and the sun passing twise a yeare by the Zenith, right ouer their heads (as when the sunne is in the beginning of Aries and Libra.) And to them also dwelling vnder the Equinoctial are two summers and two winters: and the heat is mightiest and strongest, when the sun draweth from them into the North, or South; yet doeth the sun alwaies burne the earth right vnder it, causing a burning Zone, and not parteth far from their heades. So that their winters are not perfectly and simply named winters, as with vs which are cold seasons in dæde; but rather with them is a continuall summer: yet for that the causes of heate with them, are not vnformally, and in a like maner alwaies, for that the sun doth not approch equally the Zenith of that parte, as the same is known to many: whereof the heat to them is not vnifourme and alike in burning. But sometimes hotter, and sometimes slacker and meaner of heate. So that when the sun is in the Zenith, as in the beginnings of Aries and Libra, and that they are in their high solstices; then is the heate most vehement with them, yet not without the sun, this heate can bee called mighty. But when as the sun is gone from their Zenith, which happeneth in the beginning of Cancer and Capricorne, where their low solstices are, the heate is then slacker: that is, lesser burning. So that the weaker heate hapning in the lowe solstices, may in a manner bæ named colde, in respecte of the most burning heate, hapning in the high solstices, yet it hath the nomination of winter, although no cold may bæ felt.

What

What the offices or vtilities of the Equinoctiall are.

The causes whie the skilfull practisioners tooke and vsed the Equinoctiall, with the offices which they attributed to it, and the manifold vses that it offereth, is herein declared.

1 It measureth the motion of the first and vppermost Orbe, and sheweth the same to bée drawne about by a continuall and equall swiftnesse. For that in euery equall houre, doe fiftéene of the thrée hundereth and thrée score degrées of the same arise, and so many degrées right against, set and are hidden vnder the Horizont: and that all the thrée hundreth and thrée score degrées, in 24. houres, are turned about in the appoynted times, and in their periods continually agréeing. And as the Equatour from the Poles of the worlde (about which the first mouer is drawne, and is of either side distant by equall spaces) nor the Angle, which is comprehended & fashioneth with the Horizont doth neuer change: euen so (by the same order and like motion) doeth the first heauen or moouer euidently shew it selfe to be caried about. For the Equinoctiall measureth and determineth the motion of the first mouer, in declaring his reuolution and yeare: which yeare of the first moouer, is the time of 24. houres equall. But by what meanes the auncient astronomers first found, that the Equinoctiall is drawne about in so many houres: and it is supposed they came to the knowledge thereof, by the office of some starre, (either in the Equinoctiall, or placed neare it, they perceiued the same: as that the Equinoctial from some note marked of them, did returne to it in such a

certaine space, as afore shewed.

2 The diuers motions of the Zodiack (which hapneth to it through the twart standing or lying) as a cannon or rule, doth dyrect and point out the beginnings, boundes, and time, with the which each parts or degrees of the Zodiacke arise, or doe set: and with which they touch these or those quarters of the worlde. For all the arckes of the Equatoure, are drawne by a certaine and agreeable motion continually. The parts of the Zodiacke drawne thwartly, the Equatoure doeth not varie or is distant by like spaces from the Poles of the world, nor turned about his, but the same Poles of the worlde, which doe differ by a long space from his, and drawn about by a most vnlike motion and nothing at all agreeing in it selfe: For that some parts or degrees are caried vp sooner or quicker, and others appeare slower and later. So that these vseth a more space of time in the rising slower, and those other passe vp by a shorter and quicker space. But seeing that in the Zodiack the wandring stars or Planets, doe wander continually hither and thither, and from one side of it to another: and that vnto the middle cyrcle of it or eccliptike line, the places of all the fixed stars are referred and applied: therefore cannot the times of the rising or setting of the starres, bee knowne and noted, except they shoulde be guessed and attained, by the next arks of the equinoctiall. It also declareth the equinoctialles, which are caused in those proper dayes, in which the sun hapneth to come into the equinoctiall cyrcle. For these are caused the sun being in the first degrees of *Aries* and *Libra*, in that the Zodiacke and Equinoctiall doe crosse each other in those places; whereof Manilius thus writeth.

That these signes *Aries* and *Libra* cause a right,
Throughout the earth, a like day and night.

3 It defineth and measureth the spaces, both of the naturall and artificiall dayes. And although the sun (which

drawne

of the Circles.

drawne about with the motion of the first moouer, and in the proper motion, caried forth in the meane time by force into the contrary, when as hee causeth the times of the daies and nightes, so wel as the differences of the natural daies) mooued, and runneth in the Zodiacke; yet of his motion, the day and night spaces cannot bee gatherrd, through the diuersity and vnlikenesse of the ascending or arising of diuers parts or degrees of the Zodiacke. But seeing the same motion is of all the partes of the Equatoure; therefore are the ascentions of the arcks of the Zodiacke, caried vp with the ascending of the nighest parts of the Equatour, like arising. So that both the dayes and houres, by the equall motion of these, are not founde and distinguished by the vnlike and vnequall motion of them, in that these ascentions can be, of these two cycles.

The Greekes by no meanes like of the same, in that by a stedfast order, they do mark the day and night times; therefore they parte and deuide them into equall houres, which they named times, that from the degrees of the Zodiacke they might distinguish them. For euerie fifteene parts or degrees of the Equatour in his motion and rising aboue the Horizont, doe make an houre, and euery degree foure minutes of an equall houre: so that the quarters or fifteene minutes of each degree, doe produce and cause one minute of an houre. Also they obserued the ascentions and descentions of the signes in this cycle, for that in any region or countrie, a man may knowe the length of the artificiall day and night, by hauing a sphericall instrument, and the sun placed in the East Horizont, let the note of the Equinoctiall be moued, and after the sun being turned into the West Horizont, let the note againe of the Equinoctiall be moued into the East Horizont. So that the degrees of the Equinoctiall numbred, marked with these notes, do cause an artificiall day, counting alwaies fifteene degrees of the Equinoctiall, for an equall houre. To conclude the

length

length of the artificiall day, known by subtracting the same from 24. houres, the quantity of the night remayning shall appeare howe much it is. Last the sun being entred into this cycle, doth rise in the iust East point, and setteth full West: but in the highest of summer being come to Cancer, he riseth Northeast, and setteth Northwest: at what time the noone-tide is highest. But in the shortest time of winter when the sun is come to Capricorne, hée contrariwise riseth Southeast, and is in the noonetide lowest.

4 It distinguisheth the Equinoctials and crosseth the Zodiacke thwartly wrethed and bended to it, in two opposite points, which when the sun commeth and is in it, he causeth like spaces of the day and night: and of the same, those entraunces of the sunne, are named the Equinoctiall points.

And there are two Equinoctials caused in euery yeare: as the one, the sun entring the beginning of Aries, or the spring poynt of the crossing of the Zodiacke and Equinoctiall, in the beginning of the spring, which the Latines name the equinoctiall spring, and the Grækes, *Isemerian carinen*. And the celestiall point of the same equinoctial, the Græks name the point of our equinoctiall spring. The other equinoctiall is caused, when the sun hath his beginning of Libra, in the entrance of haruest, called the equinoctiall haruest. And the celestiall point in which the sun happeneth, they name the pointe of our Equinoctiall haruest.

These points remaine not fixed in one place of heauen, but in the going before doe procéede or moue forwarde vnder the eight Orbe, and turne before the places of the fixed stars. For the point of the equinoctiall spring, that in the first yeare of Olimpias folowed the first star of Aries of the eight Sphere, 4. degrées, and 52. minutes. And in the yeare of the death of Alexander, one degrée, and 58. minutes.

The same after the beginning of the yéeres of Iulius, Cesar,

of the Circles. 77

ar, followed 4. degrees, and 50. minutes. And in the yeare of Chrisses byrth, 5. degrees and 16. minutes. In Ptholomies time, 6. degrees, and 40. minutes, it went before the same star: and in these yeares it went before that star, 27. degrees, and 35. minutes.

So that the yearely times of the Equinoctials are come backe, from the auncient time, and moue before the marked dayes by a long space: For that the Equinoctial spring which about the beginning of the yeares of Olimpias, hapned in the first or second of Aprill. In the beginning of the yeares of Cesar, in the 25. day of March. In the time of Christ our Sauiors byrth, in the 23. or 24. day of March. In Ptholomies time, in the 22. or 23. day of March. But in our time it hapneth, in the 11. or 12. day of March, and in this yeare 1570. it happeneth in the 11. day of March, and in the 11. howre before noone, on Saturday.

The Autumnall or haruest Equinoctiall, which hapned in Christ our Sauiours time, in the 23. or 24. day of September, is brought backe and come in this our time, vnto the 13. or 14. day of September, and in this yeare 1570. shall happen in the 13. day, and in the 10. houre, and 21. minutes after noone, on Wednesday.

And through this variation of the fixed stars, and Equinoctials, is caused, that the later practitioners haue found an other quantity of the yeare, contrary to the auncients. For Hipparchus and Ptholomie, haue stablished in their time the quantity of the Tropicke yeare, to bee of 365. dayes, 5. houres, 55. minutes, and 12. seconds. The Alphonsines, of 365. dayes, 5. houres, 55. minutes, and 12. seconds, Albategnius, 365. daies, 5. houres, 46. minutes and 56. seconds. Cardanus, of 365. daies, 5. houres, 48. minutes, 41. seconds, and 47. thirds. And Thebitius hath stablished the starrie yeare to be of 365. dayes 6. houres, 9. minutes, and 32. seconds, which is the space of time, in which the sun returneth vnto the same fixed star. But the

Are

Tropicke yeare, is the suns returne, after his measuring of the whole Zodiacke, vnto the Equinoctiall or solstictall point. So that by the saide pointes changed, either in the increasing or comming sooner, as hitherto hath béene obserued, is the quantity of the yeare, found to be in diuerse and sundry wise of the practisioners.

By it also is learned and knowne which stars and images celestial, are toward the North or South from it. And by it is the starrie skie deuided into two equall halfes, of which the one halfe is toward the North, and the other towarde the South. So that the denomination, so well of the Planets, as fixed stars, are thereby learned; whether they bee Southerly, or the Northerly. An other authour writeth thus of it; that it deuideth heauen into two parts, of which the one is named Northerly, of the seauen stars in the great Beare; the other Southerly, in that the sun about the South, seemeth alwaies to abide with vs in that quarter. And if the same may be knowne, which stars are named Northerly, and which Southerly: and when the Planets are named Northerly, and when Southerly. So that by this reason, all the stars and images from it, tending toward the North, to be Northerly: and from it tending toward the South, to be Southerly.

The Northerly images, in respect of the Equinoctiall, are these.

The Bull named in latine Taurus, is adorned with 33. stars, although an other writer mentioneth but of 32. Of these, 5. are in the face, and about the eies, and in the places where the hornes are described to be, are one star a piece, which make seauen in number; named Hyades in Greeke, and Succullæ in Latine, in that they stand like to the letter Y. These in the 10. 11. and 12. degrées of Taurus, hauing their latitude Southerly: of which 4. are of the third bignesse, and one brighter then the rest in the Sou
therly

of the Circles. 79

therly eie, named properly Aldebaran, of the first bignes, and of the nature of Marse. The seauen stars on the back of this signe, named Pleiades, and in Latine Virgiliæ, but in English the clustring stars: in that they stande so neare togither that they can scarcely be numbred: yet these more regarded then any of the others, in that at the appearance of them, Summer is signified; and at the setting of them, (which is sire moneths after) winter is then in entrance, like which is not shewed in the other signes. And in our time, they are in the 22. and 23. degree of Taurus, the sun ioyneth with them euery yeare, in the thirde and fourth day of May. So that after those daies, through the suns departing from them, they are knowne to arise Heliace before the sun, and then is summer entred: which in our time hapneth about the 7. 8. 9. or 10. day of may. And when the sun is come (by his course) vnto the 22. and 23. degree of Scorpio, which hapneth in our time, in the 5. and 6. day of Nouember, then is the sunne directly against Pleiades: and the sun then arising in the morning, they doe set: and aboute these daies, (as in the 5. 6. 7. 8. 9. and 10. day of Nouember) winter is entred. These as Ptholomie writeth, are of the nature of Mars and the Moone: but all the others, being some of the third and fourth, and some of the fift bignesse, are of the nature of Saturne, and a litle of Mercurie.

The signe Gemini is placed in heauen, as that betwéen them and Taurus, is that constellation Orion standing. Their headed deuided from the rest of the body, yet imbracing one the other by bodies, and doe dyrectly set with the féete, and arise together bended, as they were lying. Of which those two be the notablest, that stand in the heads: and that clear star in the head which goeth before (named Castor, and of some Appollo) hauing besides in eyther shoulder a cleare starre, in the right elbowe one, in either

knee

knée one, and in either foote one ſtar. And the other which followeth, beeing next to Cancer, hath in the heade a ſtar named Pollux, of others Hercules, on the left ſhoulder one, in the right another, and in the other partes ſundry other ſtars, to the number of 18. knowne in both. There is an other ſtar ſtanding without the forme of Gemini, going before the foote of Gemini, and following after, called Propus: and is in our time, in the 24. degrée of Gemini. Of which two are of the ſecond bigneſſe, as thoſe in the heads, but the others are of the thirde, fourth, and fifte bigneſſe. And are all of the nature of Saturne, ſauing the head going before is of the nature of Mercurie: and that in the heade following, of the nature of Mars.

The ſigne Leo looking vnto the Weſt, is placed on the body of Hydra, and not in the head, by which Cancer is nigh vnto the halfe of it, hauing the middle deuided by the ſummer cyrcle, in ſuch ſort, that vnder that Orbe hée hath the fore féet placed, ſetting and riſing with the head. Alſo he hath in the head thrée ſtars, in the nape of the neck two, in the breaſt one, in the ſpace betwéene the ſhoulders vnder the necke or behinde the necke thrée, in the middle of the taile one, in the ende of the taile another, and in the belly one cleare or bright ſtar (named the hart of the Lion) which alſo is called a royal ſtar, in that it is more about the Zodiack then the other great fixed ſtars; and for this cauſe called a ſtar of the firſt bigneſſe, although in truth, it is but a ſtar of the ſecond bigneſſe, being of the nature of Iupiter and Mars. All the ſtars which this ſigne hath (as Ptolomie writeth) are 27. Of which many are of a greate brightneſſe, as the two in the nape of the necke, of the ſecond bigneſſe: that on the heart, of the firſt bigneſſe, another on the backe, of the ſecond bigneſſe: another in the end of the taile, of the firſt bigneſſe: and all the reſt, of the third, fourth, and fift bigneſſe.

The

of the Circles. 81

THe image named the Carter or driuer of the Car, Ptolomie doeth decke with 14. starres, being all nowe in Gemini, and of the 1. 2. 3. 4. 5. and 6. bignesse, of the nature of Mars and Mercurie. Also this image named the Carter, hath a cleare starre, named the Goate standing on his left shoulder, being a starre of the first bignesse, and in our time in the 15. degree of Gemini: borowing nature of Mars and Mercurie. And that image or constellation named the Kiddes, (beeing two small stars, standing on the left hand of the Carter) are in our time, in the 12. degree of Gemini, of the fourth bignesse, and of the nature of Mars, and Mercurie.

THe image named Perseus, hath 26. stars which forme two perticuler images: of which that which is seen on his left side, is named Gorgon, or the head Algoll. And hereof it commeth that they are called the Gorgon stars. The other seene on his right side, the ancient astronomers name the Cyccle or sithe. Also Ptholomie in the description of Perseus, attributeth to the heade of Algoll (that is Medusa) as to a perticular image, foure starres. And the brighter stars of them (being in the heade of Algoll is the 12. star) is in our time, in the 19. degree, and twenty minutes of Taurus. The following star (being of the fourth bignesse) is in our time in the 18. degree of Taurus. And Ptholomie writeth, that the head of Algoll beeing of the second bignesse, is of the nature of Saturne and Iupiter: and that on the right side of Perseus, of the second bignesse, is of the nature of Saturne and Iupiter, and is in our time in 24. and 28. minutes of Taurus.

ON the head of Aries (not far from the feete of Andromeda) standeth a figure, which the Greekes (for the likenesse of the letter Delta) name Deltoton; and the Latines,

G j.

tines, for the similitude of the fourme called a Tryangle. This figure hath two equall sides, but the third not so perfect fashioned, yet easily to bee knowne; for that it shineth brighter then many other starres about it. To which the starres of Aries are a litle Southerly. And to it Ptholomie attributeth foure stars, although all other authors affirme onely three stars, except Alphonsus, which in our time are in Taurus, being of the thirde and fourth bignesse, and altogether of the nature of Mercurie.

The image of Andromeda (placed in heauen with the armes stretched abroad, and each hand bound) Ptholomie declareth it to haue 23. stars, of the thirde, fourth, and fift bignesse, and in our time are in Aries and Taurus, whose nature resembleth Venus.

This Cassiopia is figured like to a woman sitting in a chayre, with the handes lifted vp after a wayling maner; and in the turning of the world about, she is drawne with the head alwaies vpward. Ptholomie doeth number 13. stars in that image, of the 3. 4. 5. and sire bignesse, which in our time, are in the signes Aries and Taurus, and of the nature of Saturne and Venus.

Among the Astrologians onely Ptholomie and Alphonsus doe place twoe horses in heauen : or (as I may more rightly speake) the two partes of horses : of which the one is called the fore horse, or head of the horse, to which Ptholomie attributeth foure darke stars, which in our time are in Aquarius.

The other figure named of the Arabians, Alpheratz, in English the second horse, the halfe horse winged, or Pegasus, whose fore parte is described vnto the nauell : and of this, doth the greater number of authors write. Ptholomie decketh this image with 20. stars, being of the 2. 3. 4. and 5. bignesse, which in our time are in Aquaries, Pisces, and Aries, altogither hauing the qualitie or nature of Iupiter:

of the Circles. 83

piter and Mars.

The celestiall image of that fish named the Dolphine, the ancient men placed in heauen among the starres (not far from that constellation named the Eagle.) And many of the ancient astronomers, attributed but 9. stars, to this Dolphin, which are of the thirde, fourth, and sixte magnitude, and in our time be in Aquarius, retaining the nature of Saturne and Mars.

The figure named the celestiall Arrow, placed in heauen without a bowe, to which the Swanne flyeth, is neare to the North. To this Arrow doth Ptholomie atribute fiue stars, which in our time are about the end of Capricornus, being of the fourth, fift, and sixt greatnesse, and hauing the qualitie of Mars, and a litle of Venus.

The figure named the Eagle (whereon Aquarius seemeth to fly) which many affirme to be Ganimedes, Ptholomie doth deck with nine stars, of the secōd, third, fourth, and fifte bignesse, that in his time were in Sagitarius and Capricornus, and in our time are in Capricornus, which follow the qualities of Mars and Iupiter.

Many auncient authours vsed for the celestiall Harpe, the Griepe falling, which for that there is so litle thereof mentioned, shall here bee ouerpassed. But Ptholomie giueth to this celestiall Harpe, 10. stars, being all of the first third, and fourth magnitude, and in his time were in Sagitarius (except the fift and sixte starre) which then were in Capricornus, and haue the qualities of Venus and Mercurie.

The image Hercules (named of Aratus and sundry others Eugonasis) is thus placed in heauen, it maketh the Dragon appeare to haue his heade vpright, and Hercules with the right fote, the knee beeing bended or bowed, seemeth with the left fote to thrust down the right side of his

G y.   head,

head, and in his right hand holding vp a greate naile as it were to stricke, and couered on the lefte side with a Lions skin, séemeth earnestly to fight and slea the same vnarmed. This image doeth Ptholomie describe with 29. stars, and others onely 28: which in our time are all in Libra, Scorpio, and Sagitarius, and of the quality of Mercurie.

Where Aratos, Ptholomie and Alphonsus write of two manner of Crownes (as the Northerly, and Sowtherly) therefore shall first bee shewed of the Northerly Crowne, and after of the Sowtherly in their proper place. This bright constellation named the Northerly Crowne, doeth Ptholomie declare to haue 8. stars, which in his time were all in Virgo, and at this day are in Scorpio. And in the same constellation is a bright star, of the second bignesse, by the name of the whole image: of the Arabians named Alpheta of Virgill, Geor. Guosia.

The image named the Swan, the Fowle, the Hen, and the Griepe flying, doeth Ptholomie decke with 17. stars, of the seconde, thirde, fourth, and fift bignesse, and in his time were in Capricornus & Aquarius, but in our time are in Capricornus, Aquarius, and Pisces, and bee all of the nature of Venus and Mercurie.

The image named Arctipholax, or Bootes, which in English may bee named the Heardman, or rather the kéeper of the wagon, in that he séemeth to follow the wagon: that is the Northerly stars. And Plinie writing of Boots, (which he otherwise nameth Arcturus) doth affirme, that this constellation in a manner neuer riseth, but a stormy haile ensueth. Also Arcturus is a bright star not in fashion of the first bignesse, standing betwéene the legs of Bootes, as Ptholomie writeth: but Hyginus, Rufus, and others, doe place that star in the gyrdle of Bootes. This Bootes, (after Ptholomie) hath 22. stars, which in his time were in Virgo and Libra, and in our time are onely foure of the first in Virgo and all the others in Libra. But as touching the

of the Circles. 85

the natures of them, Ptholomie doth onely write hf Arcturus, which hee affirmeth to haue the nature of Iupiter and Mars.

The image named Cepheus, Ptholomie affirmeth to haue twelue stars, of the third, fourth, and fift magnitude beeing in his time in Pisces and Aries: and in our time in Pisces, Aries, and Taurus, and following the nature of Iupiter and Saturne.

The image named the celestiall Dragon, that other ancient men name the Serpent, hath (after Ptholomie) 13. starres placed ouer all: which in his time were in Libra, Scorpio, Sagitarius, Capricornus, Aquarius, Pisces, Aries, Taurus, Gemini, Cancer, Leo, and Virgo, and in our time are in these; Scorpio Sagitarius, Capricornus, Aquarius, Pisces, Aries, Taurus, Geminini, Cancer, Leo, Virgo, and Libra: being of the 3. 4. 5. and 6. bignesse. Those that shine brightest are eight, and of the thirde greatnesse, as that third star, which is on the eye; the fift star which is on the heade called Rastaben; the 24. and 25. declining vnto the North, the 29. which standeth beyonde the furthest winding, the 30. which is nere the end of the taile, and the 31. which is at the very end. And these brighter shining stars are of the nature of Saturne and Mars.

The image named the greater celestiall Beare, and of many (for the fourme of the Starres standing together) Charles-Waine. All the seauen stars, of which two bée alike, and are séene in one place, called of the auncient the two Oxen, in that they séeme equally to moue, as yoaked Oxen. The other 5. starres they imagined to fashion the wagon and the signe or image next to it, to bée Bootes, or in English the wagon driuer, which seauen stars (being the greater Beare) are drawne once about the Pole of the worlde in 24. houres, and neuer set out of sight: For one while it carieth thrée vnto the highest, and the other foure vnto to the lowest: and an other while it draweth the

G iij. foure

four vnto the highest, and bringeth the three to the lowest. This constellation named the greater Beare, doeth Ptholomie declare to haue 27. stars, which in his time were in Gemini, Cancer, and Leo, and in our time in Cancer, Leo, and Virgo, hauing all the qualitie of Mars. But here I ouerpasse all the stars of that constellation, and onely take those which form the wagō (being 7. in number) of which foure in the order of the stars of the greater Beare, beeing the 16. 17. 18. and 19. that goe before the wagon, or the whæles of the same. The 16. 17. and 19. are of the second greatnesse, but the 18. is of the third bignesse. The 25. is of the second signe (named Alioth) that is before the beame which containeth the yoake. The 26. and 27. that stand in steade of the places of the two Oxen. All these are described, in a maner after the minde of Hyginius, but Cesar the Germaine attributeth the three stars of the taile, to the beame of the wagon, and the other Starres to the whæles of the wagon.

The figure named Cynosura, the litle Beare, or lesser wagon, did the men of Syria more diligently regarde, supposing to saile the truer and surer by it: and of this thought through their first finding of the same (to haue it called after them, Phenicen. This litle Beare after Ptholomie, called the Northerly stars, or lesser wagon (as aboue said) hath seauen stars, which in Ptholomies time were in Gemini and Cancer, and in our time are in Cancer and Leo. Of these the first star, which is on the ende of the tayle, is named the Pole star, about which the first moouer is supposed to be drawne, and is of the third bignesse. The two following stars in the taile, and the two fore whæle stars, are of the fourth bignesse, and the two hinder whæle stars following, are of the second greatnesse, and these starres haue the quality of Saturne, and a litle of Venus, as Ptholomie in primo Quadri, writeth.

The Southerly images, in respect of the Equinoctiall, are

are these.

The signe Libra is a part of Scorpio, which through the magnitude of the members is deuided into two signes, of which the figure of the one they called Libra. And that part was rightly named Libra, in that when the sun is entred the beginning of that signe, the day and night is deuided a like as by an equall ballance. For the Equinoctiall harueſt like hapneth, at the entrance of the sun into Aries as the Equinoctiall spring doeth. This signe hath eight stars (which are in forme) of which one in the Southerly ballance and another in the Northerly ballance, and of the second greatnesse. But the others which either do follow or mooue before either ballance, are of the fourth and fifte bignesse. Nine others there are which bee not in forme, placed within and without the ballances. Being all of the nature of Mars and Mercurie.

The fore part of Scorpius is so hidden of the Equinoctial cyrcle, that it appeareth to stay or hold the same vp. It setteth with the heade inclined, and ariseth right vp. The signe Scorpio hath in those (which are called the Blœs) in each of them two stars, of which the fore stars are the clearer, and haue the quallity of Mars, and a parte of Saturne. It hath also in the foreheade three stars, of which the middle stars is the clearest, and of the thirde bignesse. In the space betwéene the shoulders vnder the necke, three stars. In the belly two. In the toppe of the taile fiue, with the which he is spposed to strike, two stars. In the whole the signe hath 24. stars. That one star which is named Antares (or the heart of the Scorpion) is of the second greatnes and of the nature of Iupiter and Mars. And many stars (especially those which are placed on the body) are of the 3. greatnesse, and haue the quality of Iupiter and Mars. The stars by the forehead, of the nature of Mars, and parte of Saturne. The stars on the legges and fœt, are of the fourth and fift bignesse, and haue the qualitie of Iupiter, Saturne,

G iiij.   Mars,

Mars, Mercurie, and part of Venus.

The signe Sagitarius looketh vnto the West, and is figured with the body of a Centaure, as it were shooting arowes, beginning from the feete, vnto the shoulders. It is so placed in the winter cyrcle, that his heade onely may seeme to appeare without the same Cyrcle: whose halfe Bow is deuided by the milky cyrcle. And before his feete standeth the Crowne decked with certaine stars: he seteth headlong, and ariseth straight vp. This signe hath in the head two stars, of the fourth bignes, of the nature of Mars and the Sun. In the right elbow one, and in the forehead one. In the belly one, in the left shoulder one, of the third bignesse, and quality of Iupiter and Saturne. The stars of either side the rote of the tayle, of the fift bignesse, and of the nature of Venus, and parte Saturne. In the fore knee, one star &c. This signe in the whole, hath 31. stars. Of which those on the Bowe, on the North, South, and middle part, are of the third bignesse, and of the nature of Iupiter and Mars. And two on the left fote, of the second bignesse, the one on the right ankle, of the thyrd bignesse: and that star in the right elbow, of the fourth bignesse, and hauing the quality of Iupiter and Saturne al the others are of the fourth or fift magnitude.

The signe Capricornus looketh vnto the West, and is wholy figured in the Zodiacke cyrcle. The taile with the whole body, is deuided by halfe (of the Winter cyrcle) and reacheth to the left hand of Aquarius, hee setteth headlong, and ariseth right vp: hee hath a star on the nose, and another going before the two stars in the mouth, another following them, and another Southerly of the three in the mouth: all of the first bignesse, and of the nature of Saturn, and part Mars, and Venus. A star going before the three, vnder the right eye of the fift bignesse, and of the nature of

Mars

of the Circles. 89

Mars and Mercurie. The Southerly of the thrée following behinde the horne, of the third bignesse; the Northerly of the thrée behind the horne, of the third bignesse. The Northerlier, and Southerlier of the stars in the necke, of the fift and sixt bignesse. In the neck betwéene the shoulders seauen, on the breast two, on the belly and body seauen of the fifte bignesse, and of the nature of Mars and Mercurie. The stars on the taile, of the third, fourth, and fift bignes, and of the nature of Saturne and Iupiter. In the whole hée hath 28. stars knowne, of which the two on the hornes are of the third bignesse, but all the others be of the fourth, fift, and sixt bignesse.

The auncient astronomers as Aratus, Hyginus, and others, do assigne thrée images in one constellation: as the Hydra or monstrous serpent, on whose taile they describe the Rauen to sit, & almost in the middle of the same figure, they affirme the cuppe to stand. It is a signe in the South part, hauing the heade declining vnto Cancer: the halfe of whose winding body is placed vnder Leo, but he reacheth the taile vnto the Centaure, on which the Rauen doeth sit. To this Hydra or water serpent, doeth Ptholomie giue 25. stars, being of the second, third, fourth, fift, & sixt bignesse, his beginning in Ptholomies time was in the fourtéene degrée of Cancer, but the end almost in the fourtéene degrée of Libra: and in our time the beginning is in the 4. degrée of Leo, and the end in the third degrée of Scorpius, beeing of the nature of Saturne and Venus.

The great water Cup or pitcher, doth Ptholomie decke with seauen stars, being of the fourth bignes, which in his time were in Leo and Virgo, and in our time in Virgo, and of the quality of Venus, and a litle of Mercurie.

The Rauen (after Ptholomie) hath seauen stars, being of the third, fourth, and fift bignesse, which in Ptholomies time were all in Virgo, & in our time are in Libra, hauing the quality of Saturne and Mars.

The cellestiall figure named the Aulter, doeth Aratus

place in heauen, vnder that beast called the Wolfe, neare to the South, and standing vnder the taile of Scorpius. To this figure doth Ptholomie assigne seauen stars, that in his time were in Scorpio, of the fourth and fift magnitude: but in our time are in Sagitarius, and haue the quality of Venus, and a litle of Mercurie.

The image named the Centaur, is thus described of Aratus, that the parts of this image likned to the man, do ly within the signe Scorpius; but the hinder halfe likened to the Horse, lyeth or standeth vnder the Klees. And is likened to one hauing his right hande continually open, towarde the round aulter. And as one offering sacrifice on the aulter, which sacrifice the monster holding in his right hande to offer on the aulter, they call a wilde beast. In that monster or Centaur named of Hyginus, Chiron, doth Ptholomie number 37. starres, of the first, second, third, fourth, and fift magnitude, which in his time were all in Libra, but in our time in Libra and Scorpio. The starres standing fashioned in the forme of a man, haue the quality of Venus and Mars, and those which represent the forme of a horse, are of the nature of Iupiter and Venus.

The image named the celestiall Wolfe, doeth the Centaur seeme to hold: yet it is a seuerall constellation from the other. To which Ptholomie doth assigne 19. stars, being of the thirde, fourth, and fift magnitude, that in his time were in Libra and Scorpio, and in our time are all in Scorpio.

The celestial figure named the Riuer streached from Orion, doe some name Eridanus, which otherwise Padus, some Gyon or Nylus, and some Oceanus. To this Riuer Eridanus, that commeth from the left foote of Orion, doeth Ptholomie giue 34. starres, of the first, thirde, fourth, and fifte bignesse: that in his time were in Aries and Taurus, and in our time in Aries, Taurus, and Gemini. The last star of the 34. in the rowe (of the first magnitude) hath the

qua-

of the Circles.   91

quality of Iupiter, and all the others, are of the nature of Saturne.

The long Ship (named Argo) not the whole forme of it is described or seene among the stars (in that it is deuided from the fore part vnto the mast) that may signifie to men litle to dispayre, although the Shippe happen to breake. Aratus writeth, that the fore halfe of Argo, is turned about right with the taile of the great Dogge. But in a contrary order mooued, in that the fore halfe is seene, and the other halfe hid; much like a ship rising with the swelling of the Sea, whose fore halfe is seene, and the other halfe hid, through that hinder parte darkned or hidde, and without stars. To the ship Argo doth Ptholomie ascribe 45. stars of the 1. 2. 3. 4. and 5. magnituds. The greater of these in order 44. of the first bignesse, is that star (named of the Arabians Rubail, of the Latines Canopus) which standeth at the end of the Rother stéerer of the shippe, that in Ptholomies time was in the 17. degrée, and 10. minutes of Gemini, hauing the Southerly latitude 75. degrées, and the declination Southerly 51. degrées, and 41. minutes. And in our time is almost in the 7. degrée of Cancer, hauing his latitude Southerly 75. degrées, and declination of 51. degrées, and 34. minutes. All the other stars are of the quality of Saturne and Iupiter, and were by Ptholomies time, vnto our time in Gemini, Cancer, Leo, and Virgo.

The celestiall Hare placed vnder the féete of Orion, is as hee were running before the houndes of Orion, being fained to be a hunter. To this celestiall figure doth Ptholomie assigne 12. stars, of the thirde, fourth, and fift magnitude, that in his time were in Taurus and Gemini, and in our time are all in Gemini, and haue the quality of Saturne and Mercurie.

The image named Ingula, and also Orion, lieth thwart vnder to the section of Taurus, and hath starres standing and shining before the féete of Taurus; named Orion of the
worde

worde Vrina: that is, of the floude of waters. For in the winter time (when this image or constellation ariseth) he troubleth both the Sea and Land, with showers of raine, and tempests. The Romanes also name him Ingula, for that he appeareth armed, as girded with a sword, whose shape is terrible and most cleare to be seen in the shining of the stars. For if it shineth bright and cleare, then doth it portend fayre weather to follow, if it appeare dimme, then doth it threaten a tempest to ensue. The head of this signe is drawn by three stars, of which the two cleare stars, are called the shoulders, betweene which stars the necke is imagined to be, ane thereof named Ingular. Plinie doth often make mention of Orion, as of his rising and setting whole, and in some places of part, as his gyrdle, or sword. Also he doth number Orion among the fearefull stars, causing tempests. To this Orion doth Ptholomie assigne 31. stars, which whiles hee liued, were all in Taurus and Gemini, of the 1. 2. 3. 4. 5. and 6. magnitude, and one cloudy. The second star is of the first bignesse, and the thirde is, of the second bignesse, in the order of the stars of Orion, which are in the shoulders, and haue the quality of Mars, and Mercurie. The constellation named the Zone or gyrdle of Orion, hath three stars shining very bright, of the second greatnesse, in the order of the stars of Orion, beeing the 26. 27. and 28. That figure named his sworde hath 6. stars of the third and fourth bignesse, decked in the order 29. 30. 31. 32. 33. and 34. The figure named the Clubbe that Orion bare in his right hand, when he fought with the dreadfull Bull, that possesseth foure stars, of the fifte and sixt bignesse. In the order 9. 10. 11. and 12. of these the 9. and 10. are in the right hand. Further the other stars, either of the first or second bignesse, as the 26, 27. and 28. bee of the nature of Iupiter and Saturne. But the other stars which are in the 3. 4. 5. and 6. and the cloudy star, do imitate the quality of Saturne, the 35. which is

on

of the Circles. 93

on his left foote, is of the Arabians named Rigel, of the first bignesse, and referred to the nature of Iupiter; but the others vnto the quality of Iupiter and Saturne.

The auncient astronomers placed two Dogges in heauen, as they were following the Hare running: of which the one they named Procion, and the other the Dag. The image named Proceon (in English the fore-Dogge) hath no other name with the Romanes, then the Caniculer: that is, the lesser Dog. And of Tully (*in fragmentis Arati*) hee is named the fore-Dog. But the other doeth Aratus place vnder the hinder feete of the fore-Dog. To this fore-Dog doth Ptholomie attribute onely two stars, others do number three, that in Ptholomies time were in Gemini, and in our time are in Cancer. Of which the fore star which is in the addition of the same, doth possesse the magnitude. The scond star, which standeth on the legge shining bright, is Procion; of the first bignesse, al are of the nature of Mercurie, and a litle of Mars.

The other Dogge being the greater, is named of the Arabians Alhabor, which properly is named the greater dog. And this vnderstand, that the same starre is brightest shining, which standeth on the mouth or tung of the Dogge, being of the first bignesse, and named by authors the dog, in the name of the whole image. The star named Syrius or the Dog, is placed in the middle Center of heauen, vnto which when the sun shall come, the heate is then doubled, and mens bodies affected with faintnesse. Also they suppose that star to be called Syrius, through the brightnesse of his fiery shining. The Latins name him, the Caniculer or Dog-star. Of which the Caniculer or Dog daies were named: in that whiles the sun runneth in that part, it is dangerous, and this through the quality of the season then being, that disposeth the time to health or sicknesses. And hereof it is, that whiles for a time it ariseth, the season is not alwaies contagious. Ptholomie nameth that Starre
which

which is on the mouth of the Dog; and assigneth him to be of the first bignesse, most cleare and bright in shining. And to that star which standeth or is placed on the head, he giueth a small quantity: that is, to be a star of the fift bignes. Auicen thus writing of the Dog-daies, willeth men to beware, and learne the time in which the greater Dog ariseth, and the season in which the snow lieth stil on the high hilles or mountaines, and the frosty or sharpe colde time, for then is no apt time of ministring medicine. But a medicinem:y safely bee druncke, or otherwise giuen, in the spring and haruest time. Hippocrates beeing of the same minde, affirmeth that in the Dog-daies, and before them no purgation may safely be ministred. The beginning of the Dog-daies varieth, according to the diuersity of Regions, Climates, and Latitudes. In our time the Doggedayes begin at the suns entrance into the 10. 11. and 12. degrée of Leo. That which aboue was said, that the star Syrius is in the middle Center of heauen, is ment that the star is in a celestiall cyrcle, as the Solsticiall colure, whose Center is the Center of heauen, in which that cyrcle is described. This vnderstand, that the sunne is then ioyned with the star Syrius, when they both arise together in the Horizont aboue the earth, and setteth Heliace West with the sun, though it cannot be séene rise in the morning, for the bright beames of the sun: but after the suns dayly moning from it, the star beginneth to arise and be séene in the morning before the sun. To a man of knowledge this is not strange, that the Dog-star ariseth once euery naturall day: yet the words of Auicen are thus ment, that in what time the Dog-star ascendeth with the sun, and this at the Horizontall méeting and ioyning together of them in the morning: which pestilent Caniculare time do the physicians determine to be of 40. daies long. But the malice of that season is many times ouercome & changed through the strong beames of the Planets hapning in this time: as

of

of Iupiter, Venus, and sometimes of Saturne. Ptholomie doth assigne to this constellation named the Dog-star, 18. stars, of the first, 3. 4. and 5. magnitude, that in his time were all in Gemini, and in our time in Cancer (except the 17. star) which is in the end of Gemini. And that which is brightest shining in the Dogs mouth, is named Alhabor, hauing the quality of Iupiter, and a litle of Mars, and al the others applied to Venus.

Hyginus writing of this image, named the Southerly Crowne (which of many is named Vraniscus) as if the same appeared fashioned hollow from heauen. The same doth he thus describe, that before the fore feete of Sagitarius, are a fewe stars, fashioned into a roundnesse, which forme his Crowne, that many haue imagined as cast from him in bondage maner. And many meane by this Greeke word *Ouraniscos*, the Palat, in that this crowne appeareth fashioned like to the Palate, which is a hollownesse aboue the toung. To this celestiall Crowne, fashioned like a litle Palate, doth Ptholomie assigne 13. starres, of the fourth, fift, and sixt magnitude, that in his time were all in Sagitarius, and in our time are in Sagitarius and Capricornus, of the nature of Saturne and Mars.

This image do some name the monstrous Fish, the terible fish, the monstrous sea beast, and sea Lion or Beare. This huge fish named the celestiall Whale, is placed vnder Aries, and both the fishes, lying a litle aboue the starry Riuer in the Region of heauen. Ptholomie doth assigne to this celestiall Whale 22. stars, of the third, fourth, and fift bignesse, that in his time were all placed in Pisces and Aries, and at this day are in Pisces, Aries, and Taurus, and most of them are of the nature of Saturne and Venus, and some onely of Saturne.

This Meridiane fish (named the Southern or Southerly fish) and greate, whose Aliances are the fishes named, which are placed in the cyrcle of the Zodiacke. This signe or image is placed in the South parte, and seemeth (as it were) with the mouth to drincke of the water comming from the signe Aquarius. Ptholomie doth number and giue to this Southerly fish, 11. stars, being of the first, fourth, and fift magnitude, that in his time were all in Capricornus and Aquarius, and in our time are all in Aquarius. The brighter starre in his mouth, hath the quality of Venus and Mercurie. But those stars placed on the body of the same, are agreeable and a like to the nature of Saturne.

These hitherto for the images placed on the North and South side of the Equatoure.

5. By the sift, is the declination of the parts of the Eclipticke from the Equatoure, as at the bound from which it is knowne, and both the declinations of the stars, and the latitudes of places learned. The declinations of the stars are called the distances of them from the Equatour, toward either of the Poles of the worlde. The latitudes of places, the spaces from the Equatoure vnto the highest of them raised in the Meridiane, as by the toppes gathered and learned, in the standing right ouer. Also by the Equatoure doe we learne the declinations of the Planets, aswel Northerly, as Southerly moued, as more euidently doth appeare, in the solyde Sphere or Globe. So that by the declination of the stars knowne, a man may easily place them in proper instrumentes, by which greate vtility ariseth. And it is the measure of time, in that the length of the naturall day is knowne thereby.

6 By the sixt is learned, that in the same Cyrcle (as by the subiect) is both the length of the whole earth, and perticular places standing in diuers parts of the earth, considered and measured. For according to the exact doctrine of the sphericall tryangles, the longitude or length of places, and the difference of longitudes is alwaies the Equinoctiall Arke, and not any Parallell. By it also the declination of any degree of the Zodiacke is knowne, which be-

of the Circles. 97

being had in any day at noone (the sun then shining cleare forth) the Northerly latitude or eleuation of the Pole of any Towne, may artificially be knowne. It is besides the measure of time, in that a naturall day is perfourmed by one whole returne of the Equinoctiall, with an adition or increase to that parte of the Eccllipticke, which the sun in the meane whiles accomplisheth by his proper motion, against the motion of the first mouer.

7 By the seuenth, it much auaileth and helpeth the doctrine of astrology, in that by the guide and leading of the same, are the beginnings of the twelue houses of heauen found, when astrologiall figures are erected and fashioned to prognosticate or iudge by: which can neuer so perfectly be searched and found without the Equatoure, and this through the vnlike motion and ascention of the parts or signes of the Zodiacke. By it also are all Townes according to their longitude and latitude, easily placed and found in the earthly Globe: so that by it a man may readily know which Townes are Northerly, and which Southerly. It hath besides a most great vse in Geography, vnto finding the distances of places, and vnto placing of Cities in the earthly globe, in hauing the true longitude and latitude of them.

8 The eight instruction, that by it a man may attaine the knowledge of all the celestiall Parallell cyrcles, and the earthly Zones lying vnder them. As by this example, the Parallel streached along by Rodes, cannot otherwise be knowne, but by his distance from the Equinoctial as by his principall & fore noted Parallell: which a man may learne and know to bée from the Equatoure, toward the North 36. degrées. The same knowledge may aptly be had, of all the other Parallell cyrcles rightly knowne, so that none (otherwise) can bee prompt and saillfull in Geographicall matters. Cleonedes affirmeth (prima Meteor) that it afterwards behoueth to know how to discribe

N.i.   each

each turning about of the fixed stars with the first mouer, about his Center cyrcle, as that all the Parallell cyrcles are knowne. Seeing among those cyrcles, the Equinoctiall is greatest, and those Parallell cyrcles least which are drawne about the Poles of the worlde: euen the like are those the greater cyrcles according to proportion from them, which are described vnto the Equinoctiall.

9 The niuth sheweth, that no description of the earth, (although in platefourme) can bee expressed, neither by straight nor crooked lines, without the knowledge of the Equatoure.

10 By the tenth appeareth, what commodity of the same hath and serueth in the iudging of genitures, is here by silence ouerpassed, seeing with breuity it cannot bee vttered.

The description, names, and offices of the Zodiacke, and Eccliptick line, or way of the Sunne.

After the ancient Astronomers had deuided heauen into twoe equall halfes by the Equinoctiall, and diligently obserued and noted the thwart drawing and standing of the Zodiacke, and a like forme of a larger Zone, the diuers courses, motions, and wandrings, both of the sun, moone, and other Planets which being drawne about with the first mouer, kept no equall spaces in them selues agreeing to the first moouer, nor a like distaunt in their motions from the Equatoure: but that whiles they were dayly drawn by a contrary motion of the first moouer into the East, they in the meane time wandered one whiles into the North, and another whiles

of the Circles.   99

whiles into the South, vnto a certaine elongation and distance, and so returned vnto that cycle. They obserued also that the Planets kept alwaies one maner of iourney and way, and that way cutting or crossing heauen and the Equinoctiall by a thwart manner, the same of these, they named the Zodiacke.

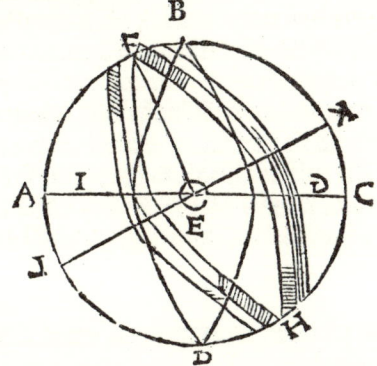

This cycle of the 12 signes, commonly called the Zodiacke (which also is a greater cycle, and thwart lying) hauing a latitude moueable vnto the motion of the sphere to which it fasteneth, and euery where is a like, vnder which the Planettes by a continuall motion are drawne and run.

This cycle also doe the Latines name thwart, through the thwart standing of it: for the Equatour doth compasse the sphere of the worlde, by the iust middle space between either Pole: but the Zodiacke is thwartly drawn both to the sphere of the worlde, and to the Equatoure: so that in some partes it is nearer to the Poles of the same, and in some parts further distance from it. It is crossed also of the Equatoure into two equall halfe cycles; of which the one is called the Boreall or Northerly halfe cycle, and the other the Meridionall or Southerly halfe cycle: therefore

H ij.   by by the continuall turning of heauen drawne about, vnto any right and thwart Horizont, inclined according to the thwart Angles, it doeth both chaunge and varie those Angles by the continuall motion and turning about. For to certaine Arks it figureth and formeth righter, and to certaine others thwarter Angels, through that diuers inclination vnto the Horizont, which ensueth after the standing of it. And the diuersitie of the inclination of it vnto the Horizont, doth also cause a varietie in the motion. For those doe flower arise, which make right Angles with the Horizont, and those are sooner drawne vp and appeare, which doe cause thwart Angles. In the thwart Sphere, (with that thwartnesse of the Sphere and the Angles, which the Horizont and Zodiack performe) is the thwartnesse encreased.

What the names are of this Circle.

1. This Cyrcle is named the Zodiacke, of this Græeke worde *zoes*, that is in English Life: in that it is the path, or the comming and going of the sun, which is called the author of life, & causer of generations (as Aristotle writeth.) Or of the Græeke name *zódion*, which in English is the figures of Beastes, with the which this cyrcle is imagined to be formed by the concourse of stars.

2. This Cyrcle is named thwart or bowing, in that it crosseth thwartly the Equinoctiall and first moouer, and doth appeare thwart in respect of the Poles of the worlde, from which it is not equally distant. Or for that it maketh not right but thwart Angles, with the Equinoctiall, and Colures, or Tropickes. Or for that it doeth not regularly

of the Circles.

ly ascend and discend according to his partes (like as the Equinoctiall doth) but that certaine parts or signes of the same doe righter and slower; and certaine thwarter and swifter arise in either Sphere. But the Zodiacke is not named thwart (compared vnto the proper Poles) seeing from them it is equidistant according to each parte; as the Equinoctiall from the Poles of the world. Yet compared vnto the Poles of the world (in that the one halfe of it declineth vnto the North, and the other vnto the South) and seeing vpon these it is thwartly drawne by the dayly motion, vnto the mouing of the Orbe in which it is: In this respect, is the circle named thwart.

3 This Cyrcle is also called Signifere, of the 12. signes caried in it, with the which the Zodiacke is described. Or for that this Cyrcle is deuided into twelue equall partes, (which are called signes) and each hauing a name of some proper beast: or for the disposing of stars being in the signe, or for some property commō to the beast and signe (which Ptholomie nameth the Orbe of the signes) Plinie, Capella, and sundry Poets doe name Signifere: but Aristotle nameth a thwart cyrcle, in that it doth thwartly crosse the Equatour, as witnesseth Proclus.

What is the cause of the thwartnesse of the Zodiacke.

1 There are two causes, why the Zodiacke is thwart: the one is, that the neather Spheres vpon the other Exe-tree, and Poles, may bee caried contrary to the motion of the first mouer.

2 The other is, that there may be diuersities of times, and varieties of qualities and temperances

pearances: that the sun also may wander and goe about diuers partes of the earth, running in the thwart cyrcle; wherof Aristotle writeth, that it is necessary, that the proper motion in the Zodiacke is vnlike to the motion of the first mouer, that it may therby cause the variety of cresent or growing things. For if there were onely one motion, there should no varietie of growing things be caused.

3 A like reason to this, that of the same, one parte of it doeth drawe nigh to the top and highest ouer our heades, and the other, that it is remoued and distant from vs, doth cause most commonly the diuersitie in effectes, which vnto the life of things is requisite. As for example, when the sun is in the Northerly halfe of this cyrcle, and neare the Zenith and highest ouer our heades, hee doeth cause a strong and mighty heat on all things of the earth, as by tryall we finde and see in the summer. If therefore the Zodiacke were not thwart, but shoulde equally approach or drawe nigh according to all the parts of it, then should the sun be alwaies a like neare vs. And when in a short time of summer he should cause such a heat, that his heat vndoubtedly should be so mighty, that nothing shoulde growe or bee increased, but that those thinges already growne vp and dryed shoulde bee consumed and burnt vp: wherefore the Zodiack is thwartly placed, that the sun mouing into the further halfe, his heat may thereby bee slaked and weakened, in which he being caried departeth from our Zenith, and cold then taken place, as apeareth in the winter. And if the sun should continually run in the South parte of the Zodiack, then through extremity of cold should all things be destroyed in the North part. And as neither heate nor cold is continuall, but successiuely, as those which bee engendred and caused by heate, and consumed by colde. So that the sun procureth (by comming nigh, and going from vs) in the Zodiacke, that it behoueth the Zodiacke to bee thwart. Also a diuersity of the Planets in the Zodiacke.

To

of the Circles.

To conclude, we see that by the comming of the sun to vs, generation is caused, & by his departure from vs, thinges wither and dry.

This cyrcle called the Zodiack (acording to longitude) is deuided into twelue parts or signes, and neither more nor fewer. And according to latitude or breadth, into 12. degrées. This cyrcle deuided into twelue signes, in that of the auncients it hath béene noted, that in euery reuolution of the sun, the moon is twelue time changed and new, and so many times hath hée full light. And that so many changes and full moones doe happen within the compasse of one yeare: by which it pleased them to deuide the Zodiacke into so many parts, according to length. But the diuision of the breadth, hath another cause; that is, of the other Planets, (except the sun) diuersly wandring from the same cyrcle. To be briefe, this whole cyrcle is deuided into 360. degrées, for the commodity of this number; in that the dayes of the yeare excéede this number by certayne partes: for the common yeare hath 365. dayes, and 6. houres.

There is a latitude atributed to the Zodiack, by which it differeth from the other cyrcles, in that they are described with one simple compasse, that it might by the larger space, containe the wandring of the Planettes, on either side the Ecclipticke line, least they should excéed the bonds. Yet the sun kéepeth one maner of way and iourney continually in the middle of the Zodiacke, and neuer declineth from it, neither vnto the right nor lefte side, but still kéeping his proper places immoueable; both in the rising and setting in either quarter, and is all alike in the winter and summer seasons. The declinations also of the sun, do shew and appeare to be equal, being on either side the Equator. So these doe witnesse, that the sun continually in his yéerly motion, describeth and kéepeth vnder that line named the Ecclipticke. But the other Planets doe neither kéepe continually the suns way, nor is drawne in a right path like him, but digressing on either side the suns way, doe wander the Zodiacke by a crooked or bending course; as one whiles moued into the North, and an otherwhiles into the South: and from thence returning vnto the sunnes way, as the like knowledge may be had and descerned by the eie.

For this cause, the learned practisioners described the suns course in the middle place of the Zodiacke, and imagined from it a latitude to bée attributed to the Zodiacke, which the auncient astronomers determined to bee of eyther side 6. degrées. But the late writers haue encreased the same, by adding twoe degrées to either side, through the digressions of Mars and Venus from the sunnes way; which hath béene obserued and noted to digresse and decline litle lesse then eight degrées. So that the latitude of the whole Zodiacke (in our time) is concluded and agréed to bee of 16. degrées, and the latitude is reached on either side, from the middle space of the suns cyrcle towarde the Poles of the Zodiacke, eight degrées.

The beginning of the longitude of the Zodiacke (although in the compasse of the circle, neither the beginning nor end can be assigned) which bendeth or is drawne perfectly round into it selfe; and both closeth and containeth it selfe: yet the practisioners haue assigned by the principall and most auncient doctrine of the godly fathers,, to bee in the poiut of the Equinoctiall spring, which is by the suns comming vnto the Equinoctiall poynte: or truer by the change of the moone that followeth nexte the Equinoctiall spring, is not to be doubted that the yeare then begun. So that they began to recken the Zodiacke from that pointe, where the motions and workings of the sun (the authour and shewer of the yearely space ensued) which after the day and night being alike, the day encreaseth, and hée ascending to vs ward, doth after abate the cold on the earth,

and

of the Circles. 105

and both slaketh and melteth the frostes and yce, and the hidden vertues againe of the earth, hee then beginneth to loose, open, chearish, and stirre vp by his liuely heate, and both looseth and sheadeth forth the dew moysture inclosed; and draweth vppe and procureth young plants to spring, through his comfortable warmth dayly shewed vpon the earth.

They deuided the whole Zodiacke according to length into twelue equall partes (which they named signes) through the moone as guide and ruler of the same: which passing yearely by the Zodiack 13. times, to the suns slower going ~~through them~~, is conioyned with him in twelue places of heauen. Those signes the ancient Greeks name *zòdia*, either by the figures of creatures, (which the fixed stars in their standing shewe and expresse) or by some naturall agreement, they so assigned names to them. Or els they appointed the names of beasts to the signes, through the congruent nature betwixt Starres and beastes. Also through the effects which the sun hath in those places. Besides these, the auncient astronomers described the other starres without the Zodiacke by images, that placed into images, they might be the commodiouser taught and expressed in heauen to the vnderstanding of yong students, and that their rising and setting might also bee the more readily demonstrated. Ptholomie named those *Dodekatemòria*, that is, the twelue parts. The Latines called them signes, and constellations. Also they named those partes signes, for that in those twelue parts, all the seasons of the yeare are noted. Again they named the parts of the signes degrees, of the dayly iourney of the sun in the Zodiack, for that in iourneying by litle and litle, he passeth through the whole Zodiacke.

They also deuided each signe into 30. parts or degrees, through the suns dayly iourneys gained of the first mouer, which in thirty dayes they declare by experience, to haue

mea-

measured and gone almost a twelfe part of the Zodiacke. Or for that the space from one coniunction vnto an other is of 30. dayes, which space (of all writers) is named a moneth. Or else in that the sunne by the same number of daies, hath measured almost this Arke or space of the Zodiacke. Whereof they named the selfe same, the thirty part of a signe, through the suns motion euery 24. houres, which the later Latines call degrees, and the Greeks *Merè*, that the ancient call parts. But the tenne partes or degrees of euery signe, the Greekes name *Dekatas*, and the Latines Faces; of which each signe doth conteyne three.

The names and characters of the signes of the Zodiacke, are these ♈ *Aries*, ♉ *Taurus*, ♊ *Gemini*, ♋ *Cancer*, ♌ *Leo*, ♍ *Virgo*. These in that they make the halfe cyrcle of the Zodiacke, declining into the North from the Equatoure, therefore doe they name them, the Boreall and Northerly signes.

The names and Characters of the other signes of the Zodiacke, are these. *Libra* ♎, *Scorpio* ♏, *Sagitarius* ♐, *Capricornus* ♑, *Aquarius* ♒, *Pisces* ♓. These in that they possesse the opposite place, and the halfe cycle reaching into the South of the Zodiacke: therefore do they name them, the Meridionall and Southerly signes.

The sunne also iourneyeth by these signes (as from the West into the East) by a contrary order to the fiust moouer, as this figure plainly demonstrateth: beginning neuerthelesse at Aries, and from Aries, passing into Taurus, and from Taurus into Gemini, and so to the ende of the signes.

They

of the Circles. 107

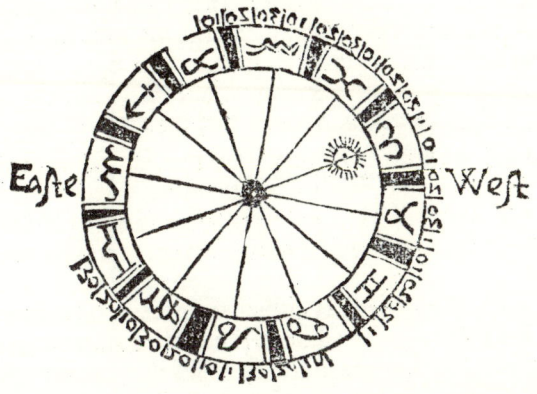

They deuided each signe, into 30. degrées of length, in that the whole Zodiacke (like as the other greater oz lesser cyzcle) containeth 360. parts, oz degrées. And as the Zodiacke hath in length 12. signes, euen so it is requisite the same should be so many degrées bzoad (as Capella wziteth. And as a degrée is in the signe the thirtieth parte oz length, the compasse of the whole Zodiacke should be the like in bzeadth. Although Mars and Venus do sometimes digresse from those bonds, yet that excesse is litle, and very seldome: and there can bée no other reason of the same, then that such a latitude is permitted oz assigned to the Zodiacke.

To this demaund, why there are onely twelue signes, and no moze, doeth Albumaser answere: affirming, that the first obseruers of the stars, noted 48. images in the 8. heauen, placed and decked with the stars, that repzesent sundzy formes, and called by them, foz the form, standing, oz nature of the stars, of which they appointed 12. foz the sunnes way: and thereforeso many, are the signes of the Zodiacke. But here may bée demanded, where the Circumference of the Zodiacke is, to which is thus answered

that

that all the circumferences of the cyrcles imagined are in that hollow of the first heauen, and likewise the signes are conceiued there to bee. And where the signes with the images of the eight sphere are moueable, and the starres in them seperated after a time. Yet the number and names, both of the signes and images remaine. So that it is not materiall, if that the starrie Aries seperate from the first Aries of the zodiacke, and the other signes the like from one another, by a most slowe course are caried, and seperated.

The auncient men deuided the partes or degrees of the signes into lesser portions, for the better attaining the precise point in the suns place. So that they appointed to ech degree 60. minutes, to each minute 60. seconds, to each second, 60. thirdes, &c. For the infinite commodity of the numbers in calculating, by reason of multiplication and diuision.

They also deuided the signes after two condicions, as in the standing, and qualities. In the standing, they distinguished them into principall, fixed, and common signes.

The principall and moueable signes, are those which nighest succeede the foure principal points of the zodiacke: of which two possesse the Equinoctiall points of the whole cyrcle (as Aries and Libra). The other twoe nighest to the Solstitiall points, are named the Tropickes (as Cancer and Capricornus.) The firme or fixed signes, next to the principall, are Taurus, Leo, Virgo, and Aquarius. The comon, or meane (or of two bodies, being the other foure) which placed (as in the middle between the principall and fixed signes) doe so obtaine a common nature of both, as Gemini, Virgo, Sagitarius, and Pisces.

In the qualities, they assigned them into foure Trients, which the Latines name Tryangles, and three cornered, the common writers nameth Triangularites, or Triplicities

of the Circles.

cities. The first trient containeth Aries, Leo, and Sagitarius, which are by the space of foure signes inclusiuely distant, or of 120. degrees: that are hot and dry, fiery, cholericke, and masculine.

The second Tyient comprehended Taurus, Virgo, and Capricornus; which beeing distaunt by the like space, are colde, and drie, earthly, melancholicke, and feminine.

The third Trient hath Libra, Gemini, and Aquarius, which beeing distaunt by the space of foure signes, are hot and moyst, sanguine, aereall, and masculine.

The fourth Trigon or Trient, doeth containe Cancer, Scorpio, and Pisces, which are distant by the space of foure signes, and are in quality colde and moyst, waterie, flegmaticke, and feminine. All which signes, are agreeing to the foure Elements, in their qualities.

Of the Eccllipticke line, or way of the Sunne.

Ere it is diligently to be considered and noted, that it behoueth not onely to know and vnderstand the places of the Planets in the Zodiacke (according to the longitude of the same) but also to learne and finde their places (according to latitude) wheather they be in that part of the zodiack which bendeth or declineth into the North, or in that parte which leaneth into the South: which the better to vnderstand and know, the ancient astronomers imagined a certaine line, going rounde about the zodiacke, and deuiding the same after length by the middle, in such sorte, that it parteth and leaueth eight degrees toward the North, & as many toward the South. So that this line is a greater cyrcle, deuiding the latitude of the zodiacke into twoe equall halfes, and hath sundrie names: as the suns way, the suns cyrcuite, the suns iourney, the suns place, the suns cycle, the Eccliptike line, and the Eccliptike place.

This line named the suns way, in that the sun keepeth alwaies the middle vnder this line, not digressing to the one side nor other: but describeth the same in his yearely motion. But the other Planets doe wander one whiles vnder it, and an other whiles on either side, which if a Planet tendeth in that part of the latitude which is vnto the North, wandring there, hee is then named to haue a latitude Northerly, as to vs dwelling Northward: but if on the other side they haue a latitude Southerly, then are they named discending and running lowe,

And by the like reason, the same line is named the suns iourney. Also of Cleomedes, called the suns cycle, in that vnder the same the sun continually runneth. And he alone being drawne by the middle of the zodiacke, neuer wandreth into the North nor South parte from that line (as we haue afore written) but continually iourneyeth about by the middle of it. So that of the same, it is called the suns cycle.

It is named the suns place, in that vnder this cycle the sun continually abideth.

To conclude, it is named by the vsuall name the Eccliptike line. For that no Ecclipse or abating of the suns or moones light hapneth, but when the sun and moone are linally vnder that line (or neere come vnto the moone) as in the same degree right against. For in the same degree, at the chaunge, is the moone come right betweene our sight, and the suns body; thereby abating his light. But the Ecclipse of the moone hapneth at the full, when as the sun is right against the moone; & that the shadowe of the earth falleth betwéen both, whereby the moons light is darkned. So that the moones Ecclipse is none other, then the falling

of

of the Circles.

of the earths shadowe betwéene the sun and moone.

The measure of the large space of either side, occupied by the Planets, limited and included by twoe lynes, and the third drawne or described by the middle, is named the Eccliptick line, and suns place.

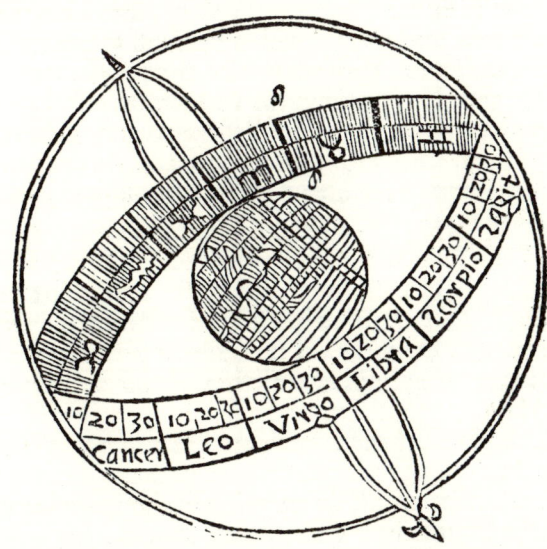

The Eccliptick line is a greater cycle consisting in the middle of the Zodiacke, and deuiding it into twoe equall compasses, defined to be eight degrées in breadth on either side, which the sunne maketh by a yearely motion, going thwartly in one continuall way, is deuided by the foure principall points; as the two Equinoctials, and the Solstices, into foure quarters. For as the whole Zodiacke, euen so the Eccliptick to the Equatour, resting as it were in two onely points, but in the rest of the cycle it bendeth from either point toward the opposite Poles of the worlde declined by the one halfe cyrcle into the North, and by the other halfe into the South.

The pointes that touch, are the Equinoctiall (as wee haue aboue written) but the other two, furthest distaunt from the Equatoure (which are as markes or boundes for the suns departure) that when he commeth to the one, hée is carried no further, but stayeth there, and from thence drawne vnto the contrary bound: through which turning backe of the sunne, they are named the Tropickes of the Græks, and of the Latines; the solsticiall points. Not for that the sun beeing carried vnto them, stayeth and remaineth any space, but neuer resteth, nor leaueth of his courses: séeing within certaine daies the Meridiane or Noone shadowes are varied, the day & night spaces either lengthened and increased, or decreased and shortned notably; as the like is yearely séene. Of these, that which in the Northerly halfe cyrcle is furthest distant from the Equatoure (named the pointe of the summer solstice) the other standing or being right against that, the point of the winter solstice.

These points change their places two maner of waies, as well according to length of the zodiacke by créeping further in the fore going, as in going to and comming short vnto the Equatoure. First that the Equinoctiall pointes, doe ouer go the places of the fixed stars, against the order and course of the signes; and therefore doe the daies of the solstices begin and goe before. For the summer solstice about the beginning of Olimpias the first day of July, which began the yeare with the Græks at the morning rising of that constellation Syrius, being notably knowne to many: but in the yeare of Christes birth, it hapned in the 24. day of June, And in the yeare 1570. it hapneth in the 12. day of June, about 11. of the clocke before Noone. The winter solstice in the first beginning of Olympias, hapned the first day of January, or there about. In the year of Christs

byrth

of the Circles.

byrth it hapned the 15. day of Decmber (in which day at the houre of 12. in the night, they affirme our sauiour to bee borne. The same winter solstice hapneth in the same yeare 1570. on the 12. day of December, aboute 2. of the clocke at after Noone.

In the second they happen vnto the Equatoure by the Eccliphcke (as it were winding) and remoued againe in the same departing. For the obseruations of many times doe witnesse, that the arcke of the Colure of the Solstices reached to these points and Equatoure (which they name the suns greatest thwartnesse or declinatio) is deminished by litle & litle. For before Pthlolomies time by forty yeeres Aristarchus, Samius, founde the same to bee of 23. degrees, 52. minutes, and 20. seconds. And Pthlolomie noted, that he found it to be iust asmuch.

Mahometes Aratensir which was after Pthlolomie 749. yeares, found this declination to be of 23. degrees, and 35. minutes.

Arzahell the Spaniard that was 190. yeares after Albategnius, found it to bee of 23. degrees, and 34. minutes.

Prophatius Iudius which was 230. yeares after Arzahell founde this declination to bee of 23. degrees, and 32. minutes.

Dominicus Maria being in the yeare of Christ 1491. found this declination to be of 23. degrees, and 29. minutes.

Vueruerus being in the yeare of Christ 1514. found this declination to be of 23. degrees, and 28. minutes, and 30. seconds.

Copernicus being a later writer, as in the year of Christ 1525. found this declination to be of 23. degrees, 28. minutes, and 2. fifts of a minute.

Of these (but by many notes considered) that the equalities haue decreased by the regulare motion, and yet shal decrease, vntill an extreame tearme of diminishing ensueth, which hee affirmeth to bee of 23. degrees, and 28. minutes:

nutes: the same Copernicus after gathered shal againe increase; and that the greatest thwartnesse which may bee caused on the sun or Eccliptike line, is 23. degrees, and 52. minutes: & the least declination to bee of 23. degrees, and 28. minutes. So that hee stablished the difference of the greatest and least to be of 24, minutes. But hee defineth the periode motion of the increasing or diminishing to be in 1717. yeares: and that so many yeares, the motion of the decrease and increase shall be, and that the whole restitution also of the thwartnesse, to be in 3434. yeares. So that as the thwartnesse failing or diminishing; euen so the points of the greatest declination (which are named the solstices) are yearely drawne and moued neerer vnto the Equatoure by 6. minutes, 27. seconds, 24. thirdes, and 9. fourthes: but dayly by one second, two thirds, and so many fourthes, caried neare vnto it. And the thwartnesse increasing, may by the like order and condicion, and in the same motion be againe abated.

As the Equinoctiall points deuide the Eccliptike lyne into a Northerly and Southerly halfe cycle, euen so the solstitiall pointes parte the same into a halfe cycle (ascending and descending) as to vs. The ascending beginneth from the beginning of Capricornus, and endeth at the last pointe of Gemini, and contayneth Capricornus, Aquarius, Pisces, Aries Taurus, and Gemini. And the descending from the beginning of Cancer, reacheth vnto the end of Sagitarius, and comprehendeth Cancer, Leo, Virgo, Libra, Scorpio, and Sagitarius. So that the sun in that halfe cycle, ascendeth from the Southerly region vnto vs, in that from vs it is digressed into the South, and of the same they receiued those names.

What.

What the latitude of a Planet is,
after two definitions.

First, that the arke of the great cyrcle is crossed betwéene the Eccliticke, and true place of the Planet, (and that is named the latitude of the Planet) for that according to the same, the Planet into latitude, that is, into the South or North, swarueth from the Eccliptick line: whereof the Planets are named to haue a latitude, one whiles into the North, and another whiles into the South. But the degrée expressed and shewed by that great cyrcle in the Eccliptick, is called the degrée of the longitude of the starre or planet, which according to longitude from the beginning of Aries vnto that place, is the Planet moued.

The other instructeth, and by demonstration sheweth, that from this line the other fiue Planettes wander, one whiles into the North, & another whiles into the South beeing not equally caried. This wandring from the saide line, is named the latitude of the Planets, and is the arke of the great cyrcle, passing by the Poles of the Zodiacke and true place of the Planet, comprehended betwéene the Eccliptick, and Center of the starre. According to this distance, he is named a Planetary starre (what starre soeuer the same be) that to latitude from the eclipticke, is carried either Northerly or southerly. So that it cannot be saide that a planet is without the Zodiacke, séeing the auncient obseruers of the stars (being moued) did attribute to this cyrcle a latitude. As may be (the Eccliptick line) noted with A. and B. and the letter C. the Pole of the Northerly Eccliptick, by which the Cyrcle noted with C. G. D.

I ij. and

116 The second Part

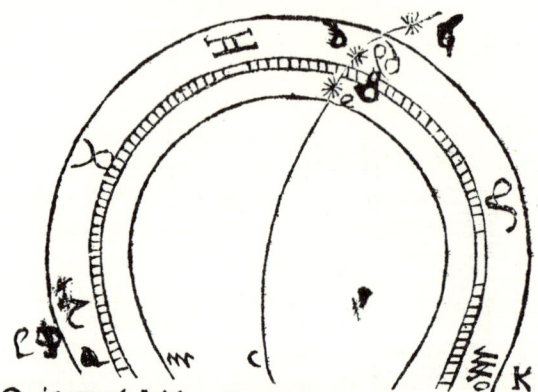

and Q. is ment, béeing the shewer of the Latitudes, and when a star shall be in the point G. he shal then be without or haue no latitude, but if in the letter E. he shal then haue a latitude Northerly, whose quantity, the arke G.E. sheweth. If so be a star shall be in the point D. then shall it be Meridional, vnto the quantity of the arke G.D. and L.D. K. the letters M. E. N. are the Parallelles drawne about the saide latitudes, on either side. So that this demonstration euidently sheweth, what the latitude of a star is; that is, when such a latitude is attributed to the Zodiack.

What the longitude of a Starre is,
and where he beginneth.

He longitude of a star, is the arke of the Zodiacke or ecclipticke line, from the beginning of Aries, reckned euen vnto that point of the ecclipticke, which is touched by the great cyrcle, drawne by the Poles of the Zodiack, and true place and degrée of the longitude of the stars. As may be conceiued in the figure aboue,

of the Circles. 117

aboue, where the pointe A. repꞃesenteth the beginning of Aries, the letters A. G. the longitude of the ſtar, if the same ſhall bee in the Eclipticke, the letters C. G. D. the cyꞃcle ending the longitudes.

The difference betwéene a declination and latitude: is this, that a latitude is the diſtance of a ſtar oꞃ Planet from the eccliptick, toward either of the Poles: which diſtance is meaſured in the greate cyꞃcle dꞃawne by the body of a ſtar, and Poles of the Zodiacke.

But the declination is a diſtance from the Equinoctial, when as the sun is caried by a continual and dayly courſe in the vpper face of the Eclipticke, and hath no latitude, but a declination onely: yet the other ſixe Planets, haue a latitude and declination. The declination of Planets, is the diſtance of them (oꞃ a degrée of the eccliptick) from the equinoctiall. And this meaſured by the cyꞃcle dꞃawne by the body of a ſtarre, oꞃ degrée of the eccliptiche, and by the Poles of the woꞃlde. The Planets alſo are said to aſcend and deſcend, by reaſon of the thwartneſſe and bending of the Zodiacke: foꞃ the sun doeth aſcend in the Noꞃtherly ſignes, but hee deſcendeth in the Southerly. In the like maner doe all the other Planets, as well by the reaſon of the aſcention, as alſo of the place. Foꞃ planets beeing in Noꞃtherly ſignes, haue the arke of aſcention greater than in the Southerly. Beſides, this part of the woꞃld which declineth into the Noꞃth, is ſuppoſed and iudged as to vs, to be raiſed higher, by reaſon of the Hoꞃizont.

Further the definition of a signe ſhall here bee declared, that the same is ment ſundꞃy wayes: one whiles to bee a circumference, an other whiles an vpper face: and ſometime to be a ſolyde body.

The Zodiacke (as I haue afoꞃe declared) is one whiles a lineall circumference (which is named the Eccliptiche line) an otherwhiles a ſwathe, of eight degrées in bꞃeadth of either ſide: Sometimes the zodiacke is called the plain

I iij.   vpper

vpper face of that Eclipticke: and in an other place the same called a solyde body, which of the saide swathe, and by the two imagined vpper faces is crossed; of which the tops or highest places ouer the head are ioyned togither in the Center of the earth, and the fœte are those Parallels of the Eclipticke ending the swathe, which may worthily be called a solide zodiacke.

The being in a signe is ment sixe waies, in that the circumference of the eclipticke (as I haue afore written) is deuided into twelue equall arks, which are called signes; and the signes ment in the first maner. Those then drawn and imagined in length by the Poles of that eclipticke line, and by the pointes of the crossings, yet by great cyrcles, as that circumference of the eclipticke: and so the zodiacke vnderstode and described of those, in the other fiue maners or waies, is deuided into twelue equall portions, whose signes are taken and ment so many waies. A signe therefore in the first maner, is mente the lineall circumference; but in the second maner the square portion of the superficiall Sphere included with foure arks: of which two are of the Parallells of the eclipticke, and two of the cyrcles deuiding, and the one ending againe in the others. A signe in the third maner, is the deuider of the circle taking here the cyrcumference, which signe is vnderstode in the first maner. A signe in the fourth maner is a certain square (pinacle wise) hauing the sharpe end turned downwarde to the Center of the worlde, and included with foure deuiders, the fœte or broader end reaching vp to the sphericall vpper face; which is a signe ment in the second maner, as this figure more plainly demonstrateth, where the letters A. B. doe represent the Poles of the Zodiacke, the letters A. D. B. and A. E. B. the Cyrcles drawn by the Poles, the letters E. and D. doe represent the twelfe part of the eclipticke; the letters G. and K. or F. and H. expresse the latitude of the zodiack; the letters A. and E. doe shewe the signe in the

of the Circles. 103

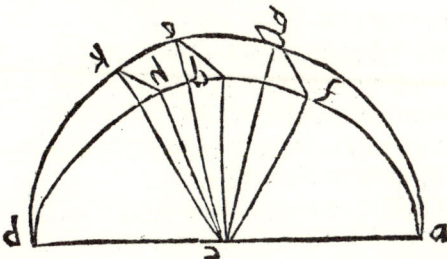

the firſt maner: the letters F. K. doe repreſent the ſquare portion in the ſecond maner: the letters C. D. E. witneſſe the deuider in the third maner: the ſharpe pinnacle (whoſe top or ende is turned downewardes) and the letter C. the foote of the ſquares: the letters F. K. do repreſent the portions in the fourth maner: the letters A. D. B. doe expreſſe a portion of the Sphericall vpper face, betwéene the halfe cyrcles. The letters A. E. B. doe ſhewe the portion in the firſt maner: the ſolide deuider betwéen thoſe halfe cyrcles is expreſſed in the ſirt maner.

Further a ſigne in the fifte maner, is a portion of the ſphericall vpper face (incloſed betwéene two halfe cyrcles) ended at the Poles of the eccliptike. To conclude, a ſigne in the ſirt maner is mente a ſolyde deuider of the Sphere, contained in the ſaide halfe Cyrcles and Sphericall vpper face, which is a ſigne vnderſtood in the ſift maner.

Theſe diuers diuiſions ſerueth (as they write) vnto that end, whereby all things might be incloſed within the ſignes. For if ſignes be deſcribed in the firſt maner, then of ſuch wiſe thoſe ſtars onely, and thoſe points are ſaid to be in the ſignes, whatſoeuer ſhall be in that cyrcumference of the eccliptike. And in the firſt ſignification alſo is ment to be vnder the eccliptike, which agréeth onely to the ſun, as at this day the ſun is (beeing the 23. day of Auguſt) in the 9. degrée, and 9. minutes, of Virgo at Noone, which is

I iiij.   ment

ment to be vnder that part of the eccliptick, that is named the 9. degrée of Virgo.

If in the second maner planets shall be in signes, which doe not excéed sixe degrées of latitude; or thus in the second signification, to be vnder the zodiacke is mente, that here a signe is expressed, inclosed within a square pinnacle portion. This signification agréeth to the other planets (except the sun) which decline from the eccliptike, as Mars in this yeare 1599. is in the 15. degrée of Virgo, which is vnder that parte of the zodiacke, that is saide to be the 15. of Virgo. Also he hath a latitude Northerly of two degrées, and 28. minutes.

If in the thirde maner, the sun or any starre shall bee in signes (placed in the plain of the eccliptick, or in the third signification to be in a signe) signifieth to be referred vnto any signe of the zodiacke. For the whole heauen is deuided into twelue Regions (in cyrcles passing by the beginnings of the signes, and Poles of the Zodiacke) of which Regions each is named a signe. And this signification agréeth to the stars, standing without the zodiacke: as the Northerly Crowne, which in our time is in Scorpio, and referred vnto the signe of the zodiacke that is called Scorpio; in that it is betwéene those twoe halfe cyrcles which passe by the beginning and end of Scorpio.

If in the fourth maner, the planets and stars also, not further distant then sixe degrées from the eccliptike. Or in the fourth signification, is ment or referred vnto any of the twelue Regiōs of heauen; into which, heauen by those sixe Cyrcles which passe by the beginninges of the signes and Poles of the zodiacke (as is afore written) is deuided. This signification agréeth to those which are in the ayre, (as be the Comets.) As if I wrote that a Comet were in Leo, here I meane the sixe cyrcles passing by the Poles of the zodiacke, and beginnings of the signes, deuiding both heauen and the whole neather Region of the worlde into

twelue

of the Circles. 121

twelue equall partes. So that a Comet is saide to bee in Leo, seeing it is in that twelfe parte which the twoe halfe cyrcles describe, running or stretching by the beginning and end of Leo.

If in the fift maner, all the stars and points ment in the vpper face of the sphere, be included in signes. If in the sirt maner, then whatsoeuer is in the world (whether the same be in the Ethereall, or Elementarie region) is accounted to be included within a signe.

Here is further to be noted, that the starres may otherwise be receiued into signes, or inclosed within signes, be-

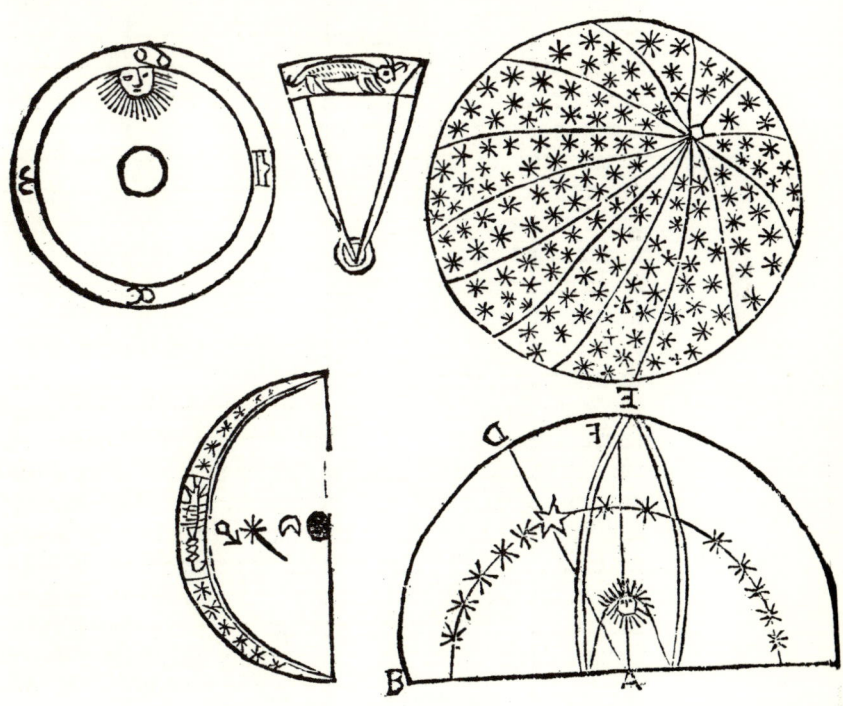

sides these sundry devisions. For the devision of the Eccliptické, alone were sufficient, that a star or any point be so brought vnto his signe, that the same may be said to bée in the signe. As by a like example the preposition (in) vsed for the worde (*sub,*) in English (vnder.) So that if the sun were saide to be in Aries, then is the same ment to bee and run vnder Aries. And for a more readier and easier conceiuing of the former lesson taught, vse the examples before demonstrated.

Here conceiue, where the cyrcumferences of the cyrcles are ment. And first imagine the circumference of the Zodiacke, and all the other cyrcles (as I haue afore written) to be in the hollow vpper face of the first mouer, and runneth as it were in the first (after the second condicion of motions) and demonstrateth alike distaunce and continuing of the cyrcles without impediment.

Although the Horizont, the Meridiane, the Verticiall, and other cyrcles (in respect of the habitation or dwelling place) remaine immoueable in that vpper face of the first mouer: yet doeth it nothing hinder (although heauen or the first mouer be drawne about) that such cyrcles bé imagined to bee immoueable. For there is nothing more agréeable, then to imagine cyrcles fixed, and those abiding in any vpper face of the earth. So that it is necessary, that the Zodiacke, the Equatoure, and the other cyrcles be described moueable in the hollow face of the first mouer, as the bound and inclosure of the whole worlde. The Horizont and Meridiane, and the others, placed immoueable in that hollowe and fixed vpper face, in which the whole earth is placed, by this meanes the fixed cyrcles shoulde stay in the fixed vpper face, and the mouable cyrcles shuld be drawne about with the moueable. As in a materiall instrument, and solide sphere a man may sée, in which the zodiacké, the Equatoure, and other moueable cyrcles, are drawne about vpon the Exe-trée, betwéene the two Cyrcles

of the Circles. 107

cles remaining steady, of which the one representeth the Horizont, the other the Meridian.

Whether the same may be described in the hollowe, or in the imbossing of the first moouer of the saide cyrcumference, it is little or nothing regarded: yet consider this, that all men may behold and sée within the heauen or first mouer, the hollowe vpper face of his inclosure, to describe and imagine the cyrcles in the same.

The Cyrcles placed without the materiall instrument, must of necessity force a man to describe the cyrcumferences of the Cyrcles in that outwarde face of the Instrument.

To conclude, the Zodiacke is ment and described according to his diffinition (being a greater Cyrcle) whose cyrcumference in the hollowe of the first moouer described into signes, degrées, and minutes, (as was afore writen) is deuided. And séeing that Cyrcle described by the suns yearely motion, is imagined straight drawne and defined or determined from the Center of the suns course by the Center of the sun, which with the sun is drawne by a perfect reuolution toward the East.

For this line in that motion cutting the hollowe vpper face of the first moner, doeth describe the cyrcumference of the Zodiacke. So that if the plaine vpper face of the suns course bée extended, vntill it cutteth the foresaide hollowe vpper face, which common section or cutting shall be the selfe same cyrcumference of the Zodiacke, vnto which the place of the force and vertue of any star is applied. Therefore by the same meanes that vertue of the moone, or any of the planets, drawne in the same hollowe of the vpper heauen, shall be like the same described.

What

What are the vses and vtilities of the Zodiacke and Ecclipticke.

The vtilities & vses that this cyrcle offereth to the studious in this art, shall here briefely be vttered.

1 It is the way, rule, and measure, of the proper motion of the planets.

2 By the benefit of this cyrcle, the true places of the stars (aswell the planettes, as fixed starres) are found. By the same also may the learned know, in which signe the planets and fixed stars are named to be.

3 This circle sheweth the latitudes of the planets and fixed Starres; the knowledge of which is greatly profitable.

4 The speciall vse of the ecclipticke is to finde out the times both of the rising and setting of the planets and fixed stars. For all are in the greatest cyrcles, being drawne by the places of the stars, and Poles of the ecclipticke and appropriated vnto the points of the ecclipticke, which those placed without the plaine of the same, doe behold toward either of the Poles. For the true places of the starres doe differ in the Ecclipticke, from them with the which they rise and set.

5 Of these cyrcles, there are certaine with arks, which lie betwéene the true places of the starres and ecclipticke, and aswell shew the true places of the starres, as their distance from the plaine of the ecclipticke; which the Gréeks call *Platos*, the Latins, latitude, in certain places in which they are caried forth, after their short course ended, and set vnder the West. Also the stars are referred vnto the Ecclipticke through the sun, which is caried in the same cyrcle,

of the Circles.

cle, and causeth the diuersity of times, and differences of daies and nights; besides, it doeth temperate by a meruelous varietie, and moderateth and ruleth the other courses.

6 By vse and experience we are taught, that vnder the ecliptick is caused the Ecclipses both of the sun and moon, of which this line came so called, for the ecclipses of the sun alwaies happen at the moones changes, and of the mone, at her opposition to the sun.

7 The thwartnesse of the Ecclipticke, as I haue afore declared, is the cause of the vnequalnesse of the artificiall dayes and nightes.

The description, names, and offices, of the Colures.

That called a Colure, is defined to bee a greater Cyrcle, turning or drawne by the Poles of the world which as vnto the motion of the Sphere is moueable: and certaine partes of it in the thwart Sphere, continually hidde vnder the Horizont.

This cyrcle termed a Colure, is so named of the Greek worde *Kolourou*, which in English signifieth mutilate and vnperfect, in that this cyrcle alwaies appeareth to vs vnperfect. But a readier knowledge of this cyrcle is, that in the thwart Sphere certaine parts of both the Colures, are nothing at all seene, in that they neuer arise to our sight, but are alwaies hid vnder the Horizont: where the other cyrcles of the Sphere (in the respect of which the thwart sphere riseth aboue the Horizont in the turning aboute of the sphere) are seene to vs in euery 24. houres. Also more or lesser of the Colures, are hidden vnder the Horizont, according to the diuers eleuation of the Pole, whereof the Colures are called vnperfect cyrcles.

There are two maner of Colures, as the Colure of the solstices, and Colure of the equinoctials. These two greater cyrcles are drawne by the Poles of the world: of which the one goeth by the Poles of the Zodiacke, and the other by the sections of the Zodiacke and Equatoure. That which passeth by the Poles of the equatour and Zodiack, doth deuide in two parts each halfe cyrcle, as well of the equatour, as the Zodiacke. Therefore the one condicion of the Colures goeth by the solsticiall pointes of the Zodiacke (which are the beginnings of Cancer and Capricorn, and the furthest pointes from the equatour) whereof it is named the solsticiall Colure. The other is named the equinoctiall Colure, seeing it entreth by the saide equinoctiall sections, which are the beginnings of Aries and Libra; that is, the equinoctiall points. So that these Colures deuide aswell the equinoctiall as the Zodiacke into foure quarters, in that they goe by the foure principall poyntes of them.

The Colures generally are called al the greater cyrcles drawne by the Poles of the world, which take their name thereof, insomuch as they neuer are descerned or seene whole in the turning about of the worlde, as the other cyrcles, but vnperfect and lacking. For both the arks right against one another about the Poles, in the thwart sphere are not seene both at once.

For they are either continually in sight to vs, and neuer drawne away or hid like vnto those which be neare to the elouated Pole. Or else they neuer appeare in sight to vs, but are continually hid from vs, as those which be the oppositcs.

But the reaching of the Colures fastneth in the two circles, extended and passing by the foure principall pointes of the eccllipticke, as the equinoctials and solsticials, which tou-

of the Circles. 127

touching one another in the Poles of the world, do in their cyrcumferences make right angles, and part the Zodiack and equatoure into foure equall quarters. The Colure of the equinoctials, resting in the equinoctiall points. The other containing the solstiticiall pointes, is called the Colure of the solstices,

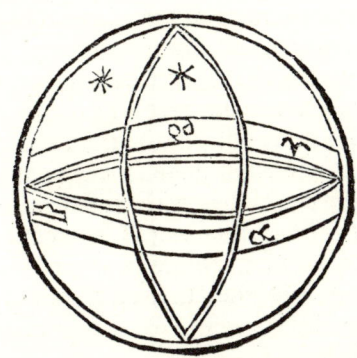

The Colure of the equinoctials is a greater cyrcle, moueable, and euery where alike : drawne by the Poles of the worlde or equatour, and the equinoctiall points, as the beginnings of Aries and Libra, making with the equatour right sphericall angles, with the Zodiack thwart angles. For oftentimes the greatest cyrcles (by a mutual touching togither) doe make right angles in the sphere, as they cut in two parts or into equall halfe cyrcles, and by the Poles one another : and contrariwaise when they cutte one another by the Poles, then doe they forme and make right angles with their cyrcumferences, and part one the other into equall parts (as writeth Theodosius in his first Booke de Sphera, and in propositi, 18. 19. 20. and 21.)

Or thus, the Colure of the equinoctials (which is named the equinoctiall Colure) is a greater Cyrcle passing by by the poles of the world, and the first pointes of Aries & Libra; where the two Equinoctiall points are said to be: in that the sunne causeth a like day and night in euery place; or for that these pointes are in the Equator, wherof it is called the Colure distinguishing the equinoctials: so that the two Colures crosse one another on the Poles of the world, at right sphericall Angles.

It is called the Colure of the equinoctials, for that it passeth by the equinoctiall pointes, as by the beginnings of Aries & Libra (which they call the Equinoctial pointes) for that when the sunne hapneth into either of them, the day and night is of equall length throughout the Earth; which commeth to passe twise in the yeare (as in the Spring and Haruest) whereof the one is called the Equinoctiall spring (and at this day is about the eleuenth of March) which is the day before Gregory: the other, the Haruest Spring, and hapneth in our time the 14. of September, that is, three daies afore Lambert, whereof ariseth this auncient verse:

    Lampert, Gregori, nox est æquata diei.

The Colure of the Solstices is a greater circle, moueable, and euery where alike drawn by the solsticiall points or the beginnings of Cancer & Capricornus, and the Poles of the Zodiacke and Equatoure, making right sphericall angles with both, for of both is the Poles comprehended. And according to Theodosius propo. lib. 2. de Spera, that by any twoe cyrcles crossing one the other (when a thirde deuideth the parts of both equally and in two partes) the same is the greater cyrcle, and passeth by the poles of both. But that which passeth by the Poles of the other Cyrcle, doth part it in two parts, and at right angles.

Here may be demanded, why the other twoe are called the solsticiall pointes, seeing the Sunne stayeth no where. Which is thus answered, that the sun digressing from either equinoctiall poynt by his proper motion, doeth dayly

de-

of the Circles.

depart from the equinoctiall cyrcle, till hée come vnto the solstitiall point, where he is furthest distant from the equatoure. But immediately after hée beginneth to returne and come againe vnto the equatour, till he come vnto the other equinoctiall pointe. So that the pointe of the suns furthest distance (which is the beginning of Cancer or Capricornus) and of the same called the solstice, in that the sun stayeth there: that is, ceaseth from his further going or departure, and beginneth againe to come to the Equatoure. For the sun after his comming vnto that point departeth, and conmeth againe to the equatour: so that the end, the departure, and beginning of his comming, is the solstice. Therefore not for that the sun stayeth there, are they called the solstices, although about those pointes of the going and comming of the sun, it is so small, that for foure, sixe, or more daies after, he séemeth in iudgement as it were to stay in one declination: and therefore for that cause may be named the solstitials. These of sundrie (as of Campanus) are also called Tropicke points, through the suns returning. And these may be called Verticiall or Cardinal cyrcles (séeing they goe by those tops of the world) and expresse or shewe the foure quarters of the Zodiacke. Moreouer séeing certaine parts of these cyrcles being neare the pole are hid, and the other right against them nothing at all discerned at any time: therefore is it that they are called in Gréeke *Kolouroi*, which is in English, maimed and vnperfect, as Proclus, Diadochus, Mocrobius and Capella, write. But this agréeth not in the right Horizont, séeing there is no part of heauen, which doeth alwaies remaine there hid. But in the description of the astrolobie, howe large soeuer the same bee, yet onely these cyrcles appeare continually vnperfect.

Yet further, the Colure of the solstices, or the cyrcle distinguishing the solstices, which also is called the solstitiall cyrcle, is a greater cyrcle drawne and passing by the poles of the world, and Zodiacke, and the greatest declinations of the same: and by the beginnings of Cancer and Capricornus. It is called the Colure of the solstices, in that it passeth by the solsticiall pointes (as by the beginnings of Cancer and Capricornus) which are named the solsticiall pointes; for that in them the solstice is caused: that is, the suns comming vnto those pointes, departeth not further from the Equinoctiall, but commeth againe vnto the Equinoctiall, which is caused twice in the yeare, as in summer and winter: whereof the one is called the summer solstice (which in our time hapneth the 12. of June or thereabout) beeing the nexte day after S. Barnabe the Apostle, where the longest day is holden to bée. The other the winter solstice, which in our time hapneth about the 11. or 12. day of December (being a day or two before Lucie) where the day is accompted shortest: whereof is this auncient verse extant.

 Vitus est Lucia, dant tibi solstitia bina.

Of the former also ensueth, that there is certaine fixed and moueable Colures. For there is a fixed Colure of the equinoctiale, which passeth by the poles of the world, and section of the equatoure and Eccllipticke of the first mouer. The fixed Colure of the solstices doth cut this at right angles in the poles of the world, and passeth by the middle of the suns greatest declination. Séeing neither the equinoctials nor solstices, are caused according to the true meaning of the astronomers (as afore may appeare) both in these points, aswell as in others: Therefore a man must conceiue, that the Colures be moueable: of which the one goeth by the true equinoctiall; that is, by the section of the suns way, and equatour, and by the Poles of the worlde, and the other of the solstices, passeth by the suns greatest declination. These hitherto written, may more plainer appeare, by this demonstration here following.

In

of the Circles. 131

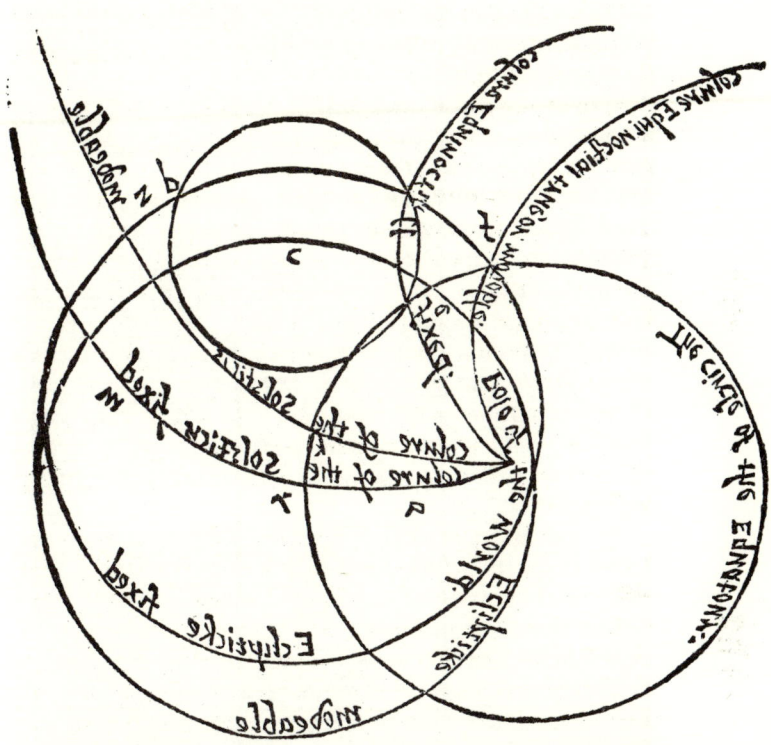

In this figure, are the cyrcles, and parts of the cyrcles noted with their names. In which the letter F. expresseth the true and moueable equinoctiall. The letter E. is a note of the fixed Equinoctiall. The letter D. pointeth out the head of Aries, of the eight sphere. The letter C. the center of the eight Sphere. The letter A. of the ninth and tenth sphere. The letters K. N. represent the suns greatest declination truely. The letters R. M. the suns middle declination greatest. How much the Equinoctials are distant

K ij   one

one from another, the former large instruction of the equinoctials, may easily shew at any time, unto which a man must alwaies resorte.

The suns greatest declination, is the arke of the Colure of the solstices, contained betwéene the equatoure and eyther Tropicke: this of sundry practisioners is diuersly noted. For Ptholomie founde the same in his time to bee of 23. degrées, 51. minutes, and 20. seconds: of which the whole cyrcle is noted to bée 360. degrées, but after Almeon, of 23. degrées, and 33. minutes. But the later practisioners haue founde the same to bee 23. degrées, and 30. minutes. Purbachius hath found the same to be of 23. and 21. minutes: which variation of the suns greatest declination is, through the comming and going of the eight Sphere, (which is named the trembling motion.) But this is demonstrated and taught more at large in the Theoricks.

If any desire to obserue the suns greatest declination, let him take the altitude of the sunne about the winter solstice, in the shortest day at Noone: which altitude beeing kept a parte, worke the like, the sun being about the summer solstice, and the suns greatest altitude at Noone found by the rule of the astrolobie, looke that in the bordure of the same, and take the middle of that arke, which is betwéene the suns least and greatest eleuation at Noone, which possesseth the myddle, and shall bee the Suns greatest declination.

The knowledge of the suns declinations, with the other stars, is very profitable; in that by the same, and the perfection of the eleuation of the pole, the true place of the sun (if the same be vnknowne) may bee knowne, the suns greatest declination presupposed, after this maner as followeth. Marcke and consider diligently (the sun being in the Noonestead cyrcle) caried vp from the Horizont: which founde, if the sun run in the Northerly signes, abate from
the

of the Circles. 133

the saide eleuation the complement of the eleuation of th Pole. If the sun bee caried in the Southerly signes, then worke contrary; for that which remaineth, shall bee the suns particular declination. As by a like example vse this. The sun beeing imagined to bee eleuated aboue the Horizont 63. degrees, 21. minutes, and 4. seconds, the eleuation of the Pole is 41. degrees, and 30. minutes, and the complement of the same altitude of the Pole to bee 48. degrees, and 30. minutes; with the which subtract the suns altitude at Noone, and the remainer shall be the suns declination, which is 14. degrees, 51. minutes, & 4. seconds, being the distance of the sun from the beginning of Aries, abated for the suns running in the Northerly signes, at the time of the obseruation before the summer solstice.

What the offices or vtilities of the Colures are.

1. Ye common offices in generall of the Colures are, to shewe the foure principall points of the zodiacke, in which through the suns motion the greatest chaunges and alterations of time is caused.

2. They serue to demonstrat the solstices and equinoctialles, and to deuide the Zodiacke into foure equall partes, to which the foure seasons of the yeare doe answere.

3. The vse of the one is to expresse and make manifest the pointes of the equinocials, and the other to shewe the points of the solstices.

4. They both cut the Zodiacke and equatour, into two equall halfe Cyrcles, and both deuide either Cyrcle into foure equall quarters.

5. But

5. But the Colure of the solsticis offereth many other vses: for in the same is the sunnes greatest declination or thwartnesse measured and numbred, in that the sunnes greatest declination, is the Arke of the Colure of the solsticis (inclosed betweene the beginning of Cancer and the Equatour) which arcke is either increased or diminished, according to the winding in and out of the eclipticke vnto the Equatour, as is afore mentioned.

6. They serue to distinguish the Equinoctiall, the Zodiacke, and all heauen into foure equall partes: the vse of which matter shall appeare in the place of the ascentions of the signes.

7. Each Colure besides, hath his priuate office or vtilities: as the Colure of the solstices, which hath foure offices. The first demonstrateth the solsticiall pointes. The second contaireth and measureth, the suns greatest declination. The third, that it stayeth vp the poles of the Zodiack, and sheweth their distance from the poles of world. The fourth, that it deuideth the Zodiack into two halfes, as into the ascending and descending. Also the same in the thwart Sphere, doeth seperate the signes rightly arising from the signes thwartly rising.

8. The Colure of the equinoctials hath twoe offices. The first, that it demonstrateth the Equinoctiall pointes. The second, that it deuideth the Zodiacke into two halfs; as into the Northerly and Southerly halfe.

9. To conclude, the Colure of the solstices doeth often supply and is vsed in the stead of the Meridiane, when as in euery dayly reuolution of the first mouer, it doth twice enter into the place of the meridiane, or is twice ioyned in the plaine of the same.

The

The descriptions, names, and offices
of the Meridiane Circles, and
Horizont.

The foure greater cyrcles which we haue already described, that with the motion of the sphere are drawn about, and euery where are alike, which the other twoe Cyrcles are contrary; as the meridiane and Horizont, that are not turned in the drawing about of the Sphere, but remaine as immoueable and fired: neither are they alike in all places, but are continually changed, standing or placed on the earth. In that all places haue their proper meridians, and Horizonts.

For both by a mutuall touching and ioyning together doe make right angles, and they continually deuide the whole heauen into foure equall parts, and make the foure angles and quarters of heauen, vnto which by a continuall turning aboute of heauen, both the one and the other stars (as it were by an orderly succession drawne) worke and send forth their vertues more effectuous, and excercise their qualities in the Elements, then in any other places: especially the sun being drawne vnto those bonds, for hee both beginneth and endeth the dayes and nightes, and distinguisheth them equally, as it were in the middle parcels of time. The sun also come vnto the meridian, doeth then more heat, dry vp and consume vapors.

The Meridiane of any place, is a greater cyrcle, which goeth or reacheth by the poles of the worlde and height of any place: and for that it passeth by the poles of the equatour, Parallels, and the Horizont, through the same doth it make right angles with them. And of this it deuideth all the arks of the Parallels as well in sight as not in sight into two equall halfes.

This circle hath sundry names, for Varro nameth it the Meridian or midday cyrcle of the nonestæd, in that when the sun is in the meridiã, or any other star aboue the earth, then hath it performed halfe the day arke, and is then at the bounde of the Nwne time. But the other halfe of the night is caused at the instant point of midnight, the halfe then reaching from East and West. So that of the same (this cyrcle of all writers in this science) named the meridiane, but of Ptholomie the cyrcle of midday and middle heauen, by the same reason.

The astrologians call this cyrcle the royall Cuspe, the regall quarter, the beginning of the tenth house, and the middle of heauen: in that this place is principall, and of worthier dignity then the other quarters, of which shall further be written in his proper place.

Further it behoueth by the addition (31. Primi Theodosij) that the Equatoure and Horizont, in the contrary maner, to passe by the poles of the meridian: & of the same that those poles is none otherwise placed, than in the common sections of the equatour and Horizont. By which sections, & by those poles of the Horizont, is a certaine thirde cyrcle greater drawne, which Iohannes (a Regiomonte) nameth the verticiall circle: so that by the foresaid Corolary or addition ensueth, that of these thrée cyrcles of each Exe-trée and Pole, are they in that common section of the other two cyrcles. Like as of the Equinoctiall, and two colures by right may be concluded. So that a triple deuision is caused by the thrée cyrcles, which appeareth on this wise: that as the meridian tendeth by our top and height from the South into the North; euen so by the same top it behoueth the other cyrcle to be drawne and passe from the east into the west, that both cutting one the other at right angle, shoulde expresse the foure foresaide quarters of the world.

As

of the Circles.   137

As the Horizont distinguisheth the upper halfe sphere, from the neather, and the meridiane from the East to the West; euen so it falleth out, that the thirde cyrcle, as that verticiall, shoulde seperate the Northerly from the Southerly halfe sphere. To these, while any standeth vpright toward the West, on such maner, that ye middle of his body is in the common Center of three cyrcles; then doth the Horizont deuide his vpper halfe from the neather, and the Meridiane, the fore part from the hinder ; and the verticall cyrcle the right part from the left. The three common sections of these cyrcles, are their Exe-trees (as is afore written) doe indicate or shewe the foure principall points of heauen (which are the sixe poles of the cyrcles) placed in the sections of the cyrcumferences, as the highest or lowest point (which are the poles of the Horizont) the point also most Easterly and Westerly, which are the Poles of the Meridiane : to conclude, the pointe most Northerly and Southerly, which are the Poles of the verticiall cyrcle.

Those people that seeme to haue their feete against ours (in respect of the roundnesse of the earth) that they dwell as it were vnder vs haue alike horizont agreeable to ours, alike meridian, and alike verticial cirzcle. But of these two, the halfe cyrcles which be extant to vs are hid or as it were vnder them. Contrariwise, those which be hid to vs, are to them extaunt. The pointe also highest to vs, is lowest to them : contrariwise the lowest to vs, is highest to them.

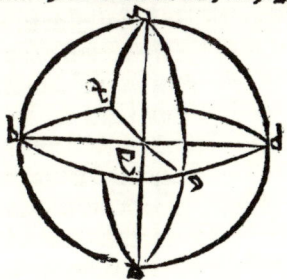

This

The figure afore placed, doth plainer expresse that afore taught: where the letters A. B. C. D. represent the Meridiane, the letters B. E. D. F. the Horizont: the letters F. A. E. C. the verticall cycle: the letter G. the center of the cycles and world: the letters A. C. the Exe-trée of the Horizont, the letters B. D. the Exe-trée of the verticiall cycle: the letters E. F. the Exe-trée of the Meridian.

To conclude, the point that to vs is most Easterly, is to them most Westerly, *et è contrario*. For the pointe most Northerly and Southerly, doe not change the surname, except you list to change or alter the names, like as of the Poles of the worlde. For that which is to vs apparant, is to them hid: and contrariwise to vs hidde, to them manifest.

Here may be demaunded whether that point of heauen most Northerly, be not the Northerly Pole of the world, and that point most Southerly, the Southerly Pole of the world. To which is thus answered, that if regard be had vnto the Equatoure and right Horizont, which passeth by the Poles of the world, and hath the equatour for the verticiall, or in stéede of the verticiall cycle. But wee which haue not the Poles of the world in the Horizont (in whose cyrcumference these foure principall pointes of the East, West, North, and South, are accustomed to be noted) are forced to call that Northerly section of the Meridiane and Horizont, the most Northerly point, and that section right against the most Southerly point. For in every place there are two sections which the meridiane and verticiall cycle doe make with the Horizont, which are two right sections in the plaine of that Horizont (cutting at right angles one an other in the Center) that expresse and shew those foure quarters of the worlde, from which the principall windes blowe; as East, West, North, and South. So that the foresaid right sections doe part the Horizont, and cyrcumference of the same into foure quarters. The foure principall

of the Circles. 139

pall windes (of the common sorte) are thus called, that which bloweth from the East, the Levant winde, and that right against it the Ponent: that from the North, Transmontanus, and that right against it the Meridional. These foure of later yeares, they haue deuided into 32. windes, after the noted lines and pointes drawne in the Saylers carde, and other Mappes euery where to be seene. Also the Saylers compasse doth expresse so many windes, directed by the adamant or lodestone, which howe the same doeth direct and shewe the windes, needeth not here be shewed, seeing the same is sufficiently known to euery sayler, which by the guide of their compasse, direct their course in clowdy weather (either by day or night) in marking diligently the points of the compasse, how they coast.

To returne vnto the matter of the Meridiane: the diuersitie of Meridianes is no otherwise caused, then the swelling of the earth, as in the first part I haue sufficiently written: the cause of which is, that one like parte of heauen cannot be the top or height of euery place. Therefore one meridiane cannot serue all places, but that in all places a proper Meridiane is caused ouer the head. The meridiane also is that which when the sun commeth vnto the highest ouer vs, foresheweth by his working and heat the midday. This meridiane is a greater cyrcle, passing by ye poles of the world and Zenith, or a direct pointe ouer the head, abiding immoueable at the motion of the sphere.

This cyrcle is differing to euery Citty and people, by reason of the East and West, and is a proper meridiane caused ouer the heade. For this is manifest, that at the chaunging of the verticiall point, there is caused an other Meridian, through the swelling and roundnesse of the earth. Also a man may of one meridian line, describe many (as writeth Iohannes a regio monte) for in that instant of the Noonetide, by letting downe right a plum line, the shadowe of the line causeth a newe Meridiane line on the

plat

platforme. Therefore these with the verticiall line in the the Center to the Horizont (crossing one an other at right angles) doe indicate the foure quarters of the worlde: as the meridian line, the North and South, the verticial line, the East and West.

The Horizont formeth the quarters of the east and west: of which the one is called the East rising, or easterly quarter or end: the other called the West setting, and quarter of the West.

The Meridiane defineth the boundes of the lowest and highest of heauen, and the quarters or middle motions of the day and night time: of which, that consisting the vpper halfe Sphere, is named the highest place and middle of heauen, the other which containeth the lowest place right against it, called the bothom or lowest of heauen.

The Meridiane is a greater cyrcle, immoueable, not one and the same euery where, but to euery place peculiar and proper, drawne by the top of the place and Poles of the worlde (vnto which the sun carried by the motion of the first moouer) doeth in the day time cause high Noone, and in the night time drawne right against it causeth mid night.

If this cyrcle were moueable like others, then at the motion of the sphere woulde it departe from our Zenith, and so lose the name of the Meridiane: neither woulde it deuide in proper place vnder it, the artificiall day into two equall parts; seeing by that motion, the Meridiane should approach neerer to one part of the Horizont, then to the other part. Nor should it stay the Horizont at right angles, of which it is numbred and accompted amongest the outwarde cyrcles of the sphere. The like affirmeth Proclus, wryting that the Meridiane is none of those cyrcles which is noted and decked with starres. For the cyrcles of the sphere are distinguished by starres, whereby those cyrcles may more easily be knowne in heauen.

of the Circles.

The meridianes are changed by the continuall chaunging of place in the swelling or imbossing of the earth (according to longitude.) For by going continually right forth toward the East and West, it doeth purchase newe Meridianes: as by going three miles forth, then is an other pointe of heauen, differing from the first ouer a mans heade and gone further by foure minutes of a degree.

Proclus affirmeth, that 300. furlongs cause no sensible alterations to happen of the Meridian: and this is ment of those which are placed vnder diuers Meridianes, and Parallelles. For those which are placed vnder one Parallell and diuers Meridianes, perceiue and see no alteration at all.

Hee which goeth strait from the North into the South, or they which directly iourney toward either of the Poles of the worlde, doe continually trauaile vnder one Meridiane. In that all Merdianes doe go from one Pole to another, therefore no iourney causeth by this meanes an other meridiane.

There are as many meridians in number, as there bee differences of verticiall points (right ouer diuers parts of the earth) in going toward the East and West. The halfe of the equinoctiall hath 180. degrees, whereof the Cosmographers doe assigne and distinguish so many meridians, in such sorte, that each Meridiane doeth passe by the twoe opposite degrees of that Equinoctiall, and Poles of the world.

For a plainer vnderstanding of the former, conceiue this figure here vnder drawne, where the letters A. C. do represent the Exe tree of the worlde, cutting the equatour and Horizont by the letter B. in the Center of the worlde; in that the plaine of the great cyrcles (when they cut one an other) that section is made in the Centers. The letters E. B. F. is the equatoure, the letters X. D. the Horizont, the letter C. the antarticke or South Pole of the worlde, the

let-

142 The second Part

letter A. the articke or North Pole of the cyrcle ending in the twoo pointes, the letters C. and A. are the Meridiane cyrcles, of these the outwarder is the meridiane fixed, as by example, passing by the fortunate Ilands, as after shal further be written in the proper place, from which the others begin, of which are commonly drawne 180. in the Cosmographicall tables or spheres. The longitude of pla-

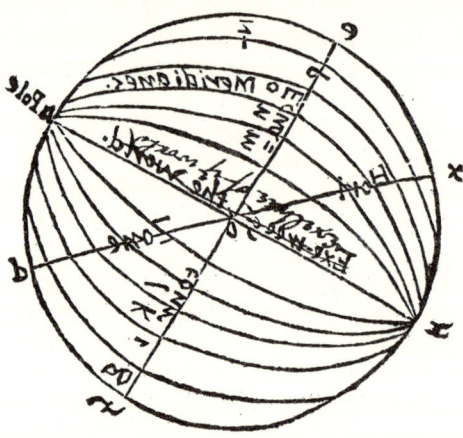

ces or citties is accompted: as for example, in the Equatour, from the point E. Westward toward the point F. in that the Cosmographers accompted the longitudes at the equatoure, & in it a degrée maketh 700. furlongs on earth. That the Cosmographers begin to accompt the longituds at the West, is for this reason; that the motions of the second mouings (that is of the planets) are accompted properly from the West to the East.

The longitude of a place is the arcke of the equinoctiall cyrcle, or of any Parallell contained betwéene two Meridianes, of which the one lyeth ouer the fortunate Iles, and the other streacheth ouer the top of the proper place noted, where

of the Circles.

where the same distance of place is gathered from the fortunate Iles at the equinoctiall, or at the Parallell of the place. The fortunate Iles are situated and lying in the sea, called Oceanus Libicus beyond Mauritania (betwéene the Equatoure and the tropicke of Cancer) which in our time is called the Iles of Canarie, and lie further into the North from the equatour, then Ptholomie noted or acounted them. But the latitude, they accompted to bee a space of the earth lying betwéene either pole, accompted in the Meridiane drawne by the poles of the worlde, or a whole tract of the earth knowne and streached beyonde, and on this side the equatoure, toward either Pole of the worlde. They stablished the beginning of the latitude in the equinoctiall (as in the middle cyrcle erquisitely betwéene either pole) and common bound to both the Southerly and Northerly places,

So that the latitude of a place, is the arke of the meridiane, betwéene the equinoctiall and Parallell drawne by the top of the place: or it is the distance of a place from the equinoctiall. This alwaies is accounted in that meridian, which hangeth directly ouer the top of the place, and to one degrée of the same, doe 500. furlongs, or 15. Germaine miles answere.

The arks of the latitudes doe not differ from the eleuations of the pole, but in the standing onely. For the eleuation of the pole is the arke of the meridian, from the Horizont vnto the Pole, raysed on high from the plaine of the Horizont. The latitude of a place, is the arke of the same meridian, placed betwéene the equinoctiall and verticiall point. To conclude, the latitude of a place, and eleuation of the pole do not differ in the magnitude or largenes, but in standing onely.

By the former figure appeareth, that the Arke of the longitude of places or citties known, is forthwith offered at the first sight: as the arke E. P. or P. O. or O. N. &c.

And

The second Part

And séeing the equatour (being in compasse about 360. degrées) doeth wholy ascend in 24. houres aboue the Horizont regularly: of this it commeth to passe (whiles in ech houre 15. degrees of the equatour doe ascend) that through the longitude of cities, it is easily knowne the hourely distance of one place vnto an other, séeing the sun commeth later to the meridian to them which are nearer to the East then to them in the West, whereof if a citty shall be situated in L. and an other in K. the arke L. K. shall be of 30. degrées: then shall the sun come sooner vnto the Easterlier meridian K. by two houres, then vnto the Westerly. But if one citty shall bee in P. and the other in Z. then (in latitude onely) shall they differ, and shall be vnder one meridian; which is declared in the last part of the description of the meridian.

What the offices and vtilities of the Meridiane are.

He vtilities and vses of this circle are many, of which the first is, that it distinguisheth the dayes and nightes into vnequal spaces: it determineth the forenoone time or morning, and the after noone or euening time of the artificiall day: the like of the night into houres (which are before night) and those which follow vnto morning. Many of the astronomers accompt their beginning of the naturall day from this cycle. It doth besides represent (without the equinoctiall) the Horizont of the right sphere, and in euery habitude of the sphere it doeth represent the right Horizont, and sheweth the points of the midday, and mid night.

2 This

of the Circles.   145

2 This cycle in the thwart sphere giueth and suprly-eth the office of the right Horizont: for to euery thwart Horizont it leaueth or stayeth at right angles. So that the astronomer maketh or accompteth not his day, from the rising or setting of the sun through the thwartnesse of the Horizont, which causeth the variety & notable difference of the inclination of the Zodiacke, vnto the horizont of the angles, and largenesse of the rising. But they begin to accompt from noone or midnight (the sun then occupying the Meridian) through the Sunne, which congruence all the meridianes haue with the right horizont. And that a lesser variety of the inclination of the Zodiacke hapneth vnto the meridiane and angles, which it maketh with the meridiane. Also in this cycle is the Zenith or direct point noted, from which the distances of the stars, and Parallel cycles are gathered.

3 The third vtility of this cycle is, that the meridian altitudes of the sun and starres are gathered and noted in this cycle, but what vtility they offer, shall sufficiently appeare in one or two examples. For when you shall haue the meridiane altitude and this in any time, then thereof you shall easily know the altitude or eleuation of the pole, if you minde to proue and try the same (the sun being vnder the meridiane of your place) take his altitude by some Instrument; as either by an astrolaby, or quadrant, which altitude found, you shall know the particular declination of the sun, by that afore taught, which shall bee caused by the suns place, at the instant time known by ye Ephemeredes. For if the sun shall be in Northerly signes, then abate the declination of the instant from the suns ascention Southerly: but if the sunne shall be in Southerly signes, then that declination shall be added to the ascention of the suns meridiane, and that which remaineth shall be the eleuation of the equatoure from the Horizont, or complement of the Latitude of the place, which is alwayes

L j.   like

The second Part

like to the eleuation of the pole. This complemente being abated from ninty degrées, you shall readily haue that which you séeke; that is, the eleuation of the pole. These by the figure here placed shall more manifest appeare, in which F.R.C. is the thwart Horizont R.B.F. the middle part of the meridiane, passing by the Zenith of the place giuen or imagined, B. the verticiall pointe or Zenith F. G. the eleuation of the pole S. B. the latitude of the same place R. T. the meridian eleuation of the sun which was sought,

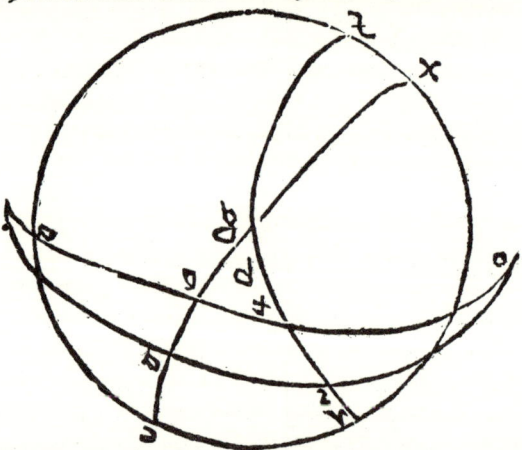

C.G.X. the Colure of the solstices, E.D. the greatest declination of the sun, A. the beginning of Aries, T. the place of the sun, A.S.D.O. the halfe of the equatoure, A.T.O. the halfe of the Zodiacke (from the beginning of Aries vnto the end of Virgo) S.T. the suns particular declination of that instant time, and G. the Northerly pole of the world. The meridiane beginneth and endeth the longitude of the earth, and particulare places on the earth, and both containeth and sheweth the differences of diuers longitudes. For the longitude of euery place from the meridiane (beginning at the fortunate Iles) endeth and resteth at the meridian streaching ouer the toppe of the same. Also it is

of the Circles.

a space included within two meridianes, of which the one resteth at the fortunate Iles, and the other ouer the top of the proper place.

4 In the meridianes (as in the subiect) the distances of the stars from the equatoure, the latitudes of places, and the eleuations of the Pole are accompted. For the studious and skillfull practisioners, obserue the latitudes of places and the eleuations of the pole, not to differ in the quantity, but in the standing onely. For the eleuation of the pole, is the arke of the meridiane from the horizont, raised vnto the pole. The latitude of the place, is the arke of the same meridian, contained betwéen the equatour and verticiall point: so that it is manifest that these arks differing in the standing, doe agrée in magnitude, whose verticiall points, one meridiane containeth, but not one Parallell, by an equall space from the west) be vnequall distant from the equatour, and are then said to differ in the latitude onely. Contrariwise, to whose tops one and the same Parallell, and not one meridian, but each place proper; those by like spaces from the equatour, be distant by vnlike spaces from the fortunate Iles, and are said to differ in the longitude onely. So that in both they are saide to differ ; to whome the Parallell only serueth, and they to whome the proper meridiane serueth: for they haue their spaces vnequall to either bound. Therfore the difference of latitude is the arke of the meridiane, contained betwéene the Parallelles of the two places, distant from the equatour. The quantity of the same is thus knowne; if from the halfe Equatour toward either pole of the places standing, the lesser latitude of the nearer, bée abated from the greater latitude more further off: if from the halfe equatour the places be deuided vnder (that the one halfe leaneth into the North and the other into the South) by the latitudes of both ioyned, whether one or both ly vnder one meridian or diuers meridians. For it forceth not in the meridian of both, that

L ij.   the

the latitudes bee ioyned togither, sæing all meridians are alike in the sphere.

The difference of the longitude, is the arke of the equinoctiall or Parallell, inclosed betwéene the meridianes of the twoe places, distaunt from the fortunate Iles, and in themselues: by which the longitude of one place excéedeth the longitude of another. The same longitude is the arcke of the equinoctial, séeing the places be vnder the equatour. For in the only longitude the places, the common Parallelles, and tops of both bended, doe differ: in that the Parallelles (from the equatour) toward the opposite quarters of the equall Parallelles (as places to which they be right ouer) doe likewise differ. The meridians (as is afore declared) are the greatest cyrcles of the sphere of the worlde, bended by the verticiall points of all places, but drawn to the equatoure (as by the Poles) of which they passe vnto right angles, and by a mutuall consent, make angles in the Poles of the world, which the arks of the equatour being placed betwéene those meridians, are measured, that by so much as a quarter of the cyrcle they bee distant from them: euen so the equatoure from his Poles, is on either part distant by a quarter of the greatest cyrcle. Those arks doe containe the difference of longitude, by which one of the meridianes is further distant into the East then the other; so that the angles vnto the Poles betwéene the meridianes, are rightly named the angles of the difference of longitude; and by the arks of the equatour, those also come into knowledge: for there is a mutuall relation betwéene the angles and arkes each one of them towards another, which doe measure the angles. The latitude of places, is the distance of the verticiall points from the equatour, gathered in the meridian. If then from the whole quarters of the meridians (which to the equatoure and Pole of the world, toward which the places decline (the equall arkes bee stretched to the latitudes, then the seates of the places
giuen,

of the Circles. 149

giuen, or the verticiall points of them shall be found. And the other arks from these points vnto the Pole (which by a mutuall section doe make an angle) the complements of the latitudes, be known by the degrées abated from 90. in the degrées of the latitudes.

For a more plaine vnderstanding of the former, conceiue this demonstration here following: where the cyrcumference of the meridian is described by the letters AB. and by the Pole of the world B. is the cyrcumference G.D. defined. To the verticiall point A. or pole of the Horizont is the cyrcumference of the Horizont E.F. drawn. Séeing that B. the pole of the equatour G.D. Therefore the arke B.D. a quarter of the greatest cyrcle, by which from D.G. the equatour B. the pole is distant. And as the letter A. is the verticiall point, E.F. is the pole of the Horizont: euen so the arke A.E. shall bee a quarter of the greatest cyrcle: and the quarters of one cyrcle are D.B. and A.E. for that cause are they equall in themselues. If therefore the same common be abated from both: that is, the middle ark betwéene A.B. which remaine, shall be the equall arks. And the arke A.D. is equall to the arke E.B. But the arke A.D. is from the verticiall pointe vnto the equatour, which is called the latitude of the place. That E.B. is the arke

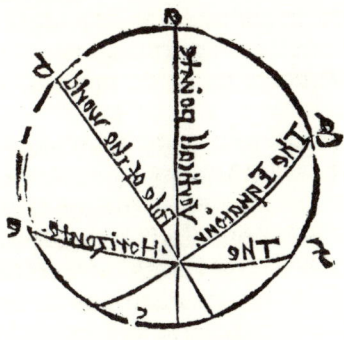

The second Part

from the Horizont vnto the pole, which is called the eleuation of the Pole. Therefore to the latitude of the place is the eleuation of the pole equall, as was afore declared.

Further by the suns meridian had and found, you may easily conceiue the eleuation of the pole, and habitude of the sphere. For the whole quarter is, of 90. degrées. Séeing the suns meridian altitude in the equinoctiall, must be subtracted from 90. degrées, the rest shew the eleuation of the Pole. As for example, the suns meridian altitude of Viteberge in Germanie, in the time of the equinoctiall, is of 38. degrées, and 10. minutes, the rest of the degrées of the quarter shall appeare to bee 51. degrées, and 50. minutes, which eleuation of the pole neer agréeth to London. So that by so many degrées, is the Pole there eleuated aboue the Horizont. And as the quadrant is from the pole vnto the equinoctial: euen so is the quadrant from the Zenith vnto the Horizont. If therefore in the time of the Equinoctiall, the distance of the Horizont vnto the suns altitude be of 38. degrées, and 10. minutes; which is not the halfe part of the quarter, the same yet being subtracted from the whole quarter, doeth shew that the rest shall bee more then halfe part of the quarter: that is, 51. degrées, and 51. minutes. For those spaces which are from the pole vnto the Equinoctiall, and from the Zenith vnto the Horizont, are alike: what the distance of the Zenith is from the equinoctiall, the same likewise is the Horizont vnto the Pole; that is, the latitude of the place, is equall to the eleuation of the pole.

To declare that the latitude of a place is equall to the eleuation of the pole, these foure propositions are to be conceiued. First, the quarters of one and the same cyrcle, any where taken, are equall one to the other. Secondly, the poles by the quarter; that is, 90. degrées bee distant from their cyrcle. Thirdly, the Zenith is the pole of the Horizont Fourthly and last, the equals abated from the equals
the

of the Circles. 151

the equals still remaine. So that two quarters of the meridian taken (as that which is from the equinoctiall vnto the pole, and that which is from the Zenith vnto the Horizont) which seeing they are quarters of one and the same cyrcle, therefore are they likewise equall one to the other; that is, either contayneth 90. degrées, when fr̄o these two quarters the common arke is abated; which is betwéene the Zenith and Pole of the worlde: and the rest of the equals remaine (as the arke which is from the equinoctiall vnto the Zenith) and called the latitude of the place; and the arke (which is from the Pole of the world vnto the Horizont) also called the eleuation of the Pole, as may be vnderstanded of the former Viteberge, that is of 51. degrées, and 50. minutes.

Yet that you may easilier finde and knowe the eleuation of the Pole of your City or Towne, you must first obtaine and haue the suns meridian altitude; which workemanly may be had and obserued by the shadow. As when the suns altitude in the time of the equinoctiall is precisely of 45. degrées, the shadowe then is like to the Gnomone, which is at Venice (as Plinie writeth) also of Milaine and Lions: for the sun to them is in the time of the equinoctial, in the middle of the quarter. But when the suns altitude excéedeth 45. degrées, then is the shadow caused lesser, as of Rome, where the sunnes meridian altitude in the equinoctiall is of 42. degrées, and 10. minutes: so that the shadowe is there shorter. Also Plinie writeth of Rome, that the ninth part of the Gnomon in the equinoctiall, doth lack of the noone shadow. But when the suns altitude is lesser then 45. degrées, the shadow of the Gnomon is caused longer. The like is with vs through all winter and the time of the equinoctiall: for we sée the shadowes of mens bodies to be longer, for that the suns altitude in that time is neuer 45. degrées. For how much the shadow is longer then the halfe part of the quarter, so much the lesser is the suns

L iiij. al-

altitude then 45. degrées. As of Viteberge in the 10. day of September, the suns meridian altitude is then of 39. degrées, and 21. minutes: but when the sun is further distant by the 45 degrée of the quarter (or by the halfe of the quarter) then ensueth, that the shadow is so much longer then the Gnomon, or 45, degrées. For the None shadow in the 10. day of September is the like vnto the Gnomon, as the 50. degrées, and 39. minutes, are vnto 45. degrées.

Heere you sée how by the meridian shadowe, you may finde the suns altitude; which obtained, you shall easily find the altitude or eleuation of the Pole (especially in the time of the equinoctiall.) For the suns altitude then from the whole quarter; that is, from 90. degrées, must be subtracted, and the eleuation of the pole shall remaine and appeare to be, as is aboue taught.

Here by the way shall bée taught how to describe and find the meridian line, whose vse is great both for the Mathematician and Architecter, in making of dials, and other necessary Instruments. To know and do this, haue a plaine body well polished and smothe, standing euen on

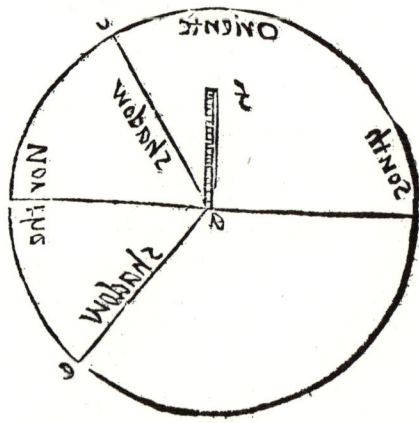

euery

of the Circles. 153

euery side, that the one side bee no higher or more leaning then the other; in the Center of which, howe large so euer the cyrcle about be drawne, set vpright a stéele, vpon pin, or other Gnomon, as here the letters A. E. doe expresse: which pin must bee very straight, and not excéede the cyrcumference of the cyrcle, but his top to be equall to it: then shall it be alike distant round about, when it shall be alike distant from thrée points at the least of the cyrcumference. But the height of the stéele or Gnomon may not excéede a quarter of the diameter of the cyrcumference, and that for the same cause, that the meridiane shadowe (which of the shadowes is then most shor)tthat falleth within the aboue said cyrcle described. In this cyrcle thus drawne and prepared, appeareth a shadow by the Gnomon of the suns shining in the forenoone, vntill it touch precisely the cyrcumference of the cyrcle, like as the shadowe A. C. demonstrateth; the point touched is noted and expressed by the point C. In the same maner is the afternoone shadow examined and found; whose point touched is also noted by the point E. So that both the shadowes end within the cyrcumference, and at the bounds also of the shadowes, are the two pointes noted. The arke contained and included by the pointes C. E. betwéene these, is deuided into twoe equall parts in the point D, by which and the Center of the cyrcle A. is the right line B. A. D. drawne, that shall bee the meridian line, which is as the common section of that Horizont and meridian. Thus is the way and rule of the meridian line described. For so Vitruuius writeth, and Iohannes a regio monte in his Capindarie, which as soone as that shadow of the Gnomon falleth on his line, it shall then be the point of noone wheresoeuer the sun at that time is placed. If another certaine line shal cut this by the Center of the cyrcle (lately described) at right angles that sha'l bee the common verticall section of the cyrcle with the Horizont, which may be named the verticiall line.

    Séeing it is somewhat harde to finde the height of the
Pole

Pole vnto any day prescribed, & that the same may more easily and surer be attained and founde, you shall vse this table here following: by the helpe of which you may without great labour, finde and know the eleuation of the pole. For to procæde and worke by this manner, sæke first the suns meridian altitude at the day offered, either by an astrolaby or quadrant; but rather by the instrument named the quadrant, in whose bordure are 90. degrées drown or written, expressed by reason of the Gnomon and shadowe vpwarde. After sæke the degrée of the Eccliptike by the Ephemerides, which the sun obtaineth at noone of the day offered: next by the table folowing, take the declination of the degrée founde (by meane of the equinoctiall) if the sun then shall bee in Northerly signes, abate or subtract from the suns altitude afore found: but if in Southerly signes, then adde vnto the suns altitude. The produce or rest is the eleuation of the equinoctiall, which abstracted or abated from the whole quarter; that is, from 90. degrées, leaueth & sheweth the eleuation of the pole, as in the 10. day of September, the suns altitude in the twelfe houre (or at noone) is of 39. degrées, and 21. minutes. To finde this eleuation of the pole, I enter the table following, where I finde and sée the 27. degrée of Virgo, to haue the declination of one degrée, and 11. minutes: which degrée and minutes (séeing they are in the Northerly part of the worlde) are to be subtracted frō the suns altitude that day, and the degrées which remaine are 38. and 10. minutes. The altitude of the equinoctial that day, which subtrated or abated from the whole quarter; that is, from 90. degrées, the eleuation of the pole which remaineth, is 51. degrées, and 50. minutes.

This

of the Circles.

This Table of the Suns declinations, containeth the number of the degrees of the Zodiacke, increasing in descending on the left hand, and increasing by ascending on the right hand, with the Signes decently placed : the Arks or roots of the declinations follow those numbers : which rootes are no other then the arkes of the circle of the Latitude : that is, the circle passing by the Poles of the Ecliptike, included betweene the Ecliptike and Equatour.

The generall Table of the Declinations.

♈	Arkes	♉	Arkes	♊	Arkes	♋	
G	Dr. m	Dr. m	Dr. m	G
0	0 0	0 12	16 20	38 30
1	0 26	12 37	20 40	0 29
2	0 12		21	28
3	1 18	12 58	21 11	27
4	1 44		21 21	26
5	2 10	13 19	21 31	25
6	2 36	13 17 40	21 40	24
7	3 2	14 0	21 49	23
8	3 28	14 20	21 58	22
9	3 53	14 40	22 6	21
10	4 19	14 50	22 14	20
11	4 45	15 18	22 21	19
12	5 10	15 37	22 28	18
13	5 35	15 55	22 35	17
14	6 0	16 3	22 41	16
15	6 25	17 31	22 47	15
16	6 50	17 48	23 52	14
17	7 15	17 5	23 57	13
18	7 39	17 22	23 2	12
19	8 3	18 38	23 7	11
20	8 27	18 54	23 11	10
21	8 51	18 0	23 15	9
22	9 15	18 25	23 18	8
23	9 39	19 40	23 21	7
24	10 2	19 55	23 23	6
25	10 25	19 9	23 25	5
26	10 48	19 23	23 27	4
27	11 10	20 36	23 28	3
28	11 32	20 49	23 20	2
29	11 54	20 36	23 30	1
30	12 16	20 30	23 30	0
♓	♍	♓	♐	♑	♒	♑

5 He meridianes with the Horizont, in any right or thwart, & in the other foure greater cycles, doe distinguish all heauen into twelue spaces, which they call the houses of heauen.

Of these foure, which occupy the angles of heauen, are called the quarters: the foure nexte to these, are named the succedents: the last (included by the succedentes and angles) are named the declining houses, and the cadent from the angles.

The meridian also hath a most great vse in Cosmographie: for by it the describers of the world measure the longitudes and latitudes of places and cities: which beeing knowne, the distance of cities may easily be found. That you may vnderstand what the longitude and latitude of a place is, it behoueth you to know the distinctiō of the earth after the Geographers, which is on this wise,

The Geographers doe assigne or imagine two points on the earth, right vnder the poles of the world: after that they deuise a cyrcle equally distant on either side frō these these two points (right vnder the equinoctiall) which deuideth the whole Globe of the earth and water into twoe equall halfes. This cyrcle thus described on earth, they distribute into 360. parts or degrées, in procéeding from the West into the East, by each degrées of this cyrcle; and by the points right vnder the poles, they imagine and draw 180. cyrcles; which, for that they are vnder the celestiall meridians, they also call meridians, and those they deuide into thrée hundred & thréescore parts or degrées; by which parts they imagine and draw the Parallell or equidistant cyrcles to the equinoctiall, procéeding from the equinoctiall on either side, towardes the pointes in the poles lying

vnder

of the Circles.

vnder these Parallels, & although they bee not of the same bignesse or largenesse (for how much nearer the poles they are, so much the narrower and strayter they run togither. Contrariwise, how farre of they bee from the Poles, and nearer to the Equinoctiall, so much the wider and larger they runne) yet doe they deuide as the Equinoctiall, or any other greate cyrcle, into three hundreth and three score partes or degrees. Nowe this deuision of the earth being learned and vnderstoode, a man may the more easily conceiue what the longitude and latitude of places is.

The longitude of a place (as I haue afore written) is the arke of the equinoctiall cyrcle, or Parallell, passing by the Zenith of the place which is sought after, included betweene the two meridians; as betweene the first meridian (which by the Zenith of the Iles of Canarie (and further off) is imagined to bee drawne, and the Meridian of the place offered: that is, the longitude of any place, is the distance thereof from that westerly point, from which the beginning of lōgitudes is accounted toward the East. They began to account the longitude from the west, through the proper motions of the Planets, which are caried vnto the contrary quarter from the West: or rather for the Moone, at whose Ecclipses it is well knowne that it more auaileth, then the true finding out the longitudes of places, or as some rather thinke likelier, that the places which ende and stand furthest Westward inhabited, haue bene surer and perfecter found. For through the nearenesse and opportunity of the iourneyes (which they in auncient time were mooued to trauaile and saile vnto) as the twoe Iles (named Gades) which lie by the furthest parts of Spaine beyond Granade, and since through the passage by West Oceean, men of later yeares haue sailed about the furthest partes without stop or impediment. But vnto the Eastward, they were stopped of their course by a great distance through the difficulty and perill of the iourney. And since

be-

beyond the halfe circle almost threescore degrees, men haue sailed to Scythia besides Imaus (which nowe is named the great Tartaria) that reacheth bordering to the vpper India, where the most large kingdome of Cathagia vnder the parallell of Thracia flourisheth, where Bebeid Cham was gouernour. And that is the part of Tartaria, which beginneth from the riuer Tanais, so that the largenesse of Scithia Asiatica (from the West to the East) doeth almost take vp 84. whole degrees. America in the sea Atlanticus, is of such greatnesse, that the same is supposed to be a fourth part of the world inhabited: the middle or halfe of it hath the longitude of 330. degrees, and the latitude of tenne degrees Southward. The sea Altanticus hath many large Ilands in it; among which, the most notable are Spagnolla, Cuba, Parias (otherwise Chersouesus) by the straight that reacheth vpward into the north. The middle of the same hath the longitude of 285. degrees: the latitude Northerly 44. degrees. For from 11. vnto 50. almost, it reacheth vnto. America streacheth far into the South, beyond the tropick of Capticorne, although his bounde or furthest part Southerly bee not yet founde or knowne. To the auncient it was no further knowne Southward, then 17. degrees beyond the Equinoctiall: and the furthest knowne to them Northwarde, exceeded not three score and three degrees, which (as Ptholomie witnesseth) was vnto the Iland Thylen. So that the whole latitude found by them, appeareth to be 80. degrees, both of the one and the other side of the equinoctial: and on earth the same containeth 40. thousand furlongs, to which 50. hundred paces answere, but Germaine miles two hundreth thousand agree. Also the Iland Thilen or Thulen, standeth beyond Scotland, and the Iles Hebrides and Orchades, that be into the North and East, which is distant from the furthest bound of Scotland, but three dayes sayling, if prosperous windes bee their helpe. At this day men haue found beyond Tnylen (but somwhat into

of the Circles. 159

into the East (and most large bounds stretched and found beyond the articke or Northerly cyrcle, & these are whole without breaking of any sea betwéene; and containe Suetia, Norway, Iseland, Grunland and Lapeland. The kingdome of Suetia appeareth most large, and containeth sundry nations and people; among which, they are of most account, the East and West Gutland people, inhabiting neare to Norway.

And vnder the king of Suetia are the Lapeland people, (as the Finelapons and Dikilapons) where are a wild and fierce people, dwelling almost vnder the pole articke (especially the Lapeland people) to whome the sun neuer setteth in the summer for 40. dayes space. Aboue these inhabit a people of a cubite long or high, hauing small and crooked bodies (named of some Pigmalions) that liue vnder a very darke and bitter cold ayre or sky. And aboue Scania (néere to the West boundes of Suetia) doeth Norway stretch into the North, whose vttermost limit extendeth vnto the 71. degrée almost of the Northerly latitude. Aboue this is the country named Iseland, by reason of the frozen waters and sea: where throughout the yeare it so bitterly frǽzeth, that through the ycie seas there thicke frozen, it permitteth no ships to come vnto thé, except in the thrée hottest months of the yeare. It aboundeth with brimstone, and burneth in many places through the sulphure & brimstone veines. Plinie writeth, that the Occean sea in the North is very large, which in these our dayes is well knowne. This also was learned of certaine skillfull sailers (which inhabited and very much had traualed this coast) that they knew not the limits or bounds of this sea toward the North, but supposed that this sea did compasse the whole earth. By this sea dwell many and mighty people; as the Danes, the Swedens, Norwaies, Gotelandes, Finelands, Russians, and Pruthenians: and vnder the pole artick the Laplands. The reason why in these places such force of moysture aboundeth,

deth, is for that a dayly and continuall cold of these places gathereth and thickneth the ayre, and by a continual working resolueth into water. For when the ayre is not throughly purged by the suns beames, then the weaknes of them, and far distance of the sun from these places, must of necessity bee continually thicke and darke, which afterwardes yældeth and giueth plentifull floodes by deawes and raines. Albert mag. in his booke de natura loci, and 8. chapter, assigneth a witty and laudable reason, why the Northerly be inhabitable. The cause he setteth downe, in that sundry skillfull Mariners affirme (that haue many times sailed into the Northerly partes of the Occean sea) that in those places is a continuall darknesse, which when men sawe they returned for feare, supposing (nay rather doubting) that none coulde saile any further in that quarter of the worlde, through the darknesse, and thicke mist, which hindreth the direction of their iourney. So that the nature of those places cannot bee sufficiently knowne to vs, seeing no man (as the learned report) hath attempted thither, through extremitie of colde their bearing sway. And for that exceeding cold is a mortifying quality, therefore a man may coniecture, that few liuing creatures and beasts can there liue &c. Yet the part of the Northerly Occean (vnto the Easterly side) is sufficiently knowne to many trauailers.

Although the vttermost boundes of the earth are not wholy knowne, yet the nearest approaching to them shall here bee applied, as the longitude of the earth distaunt betwæne Peru (the Realme of America) and Cathaya, to expresse 315. degræs: or if any minde to accompt the longitude from the fortunate Iles, they may by a whole cyrcle containe them, euen as the whole Orbe about in a maner doth partly giue place to the water, and are partly dwellings for men, beasts, and other liuing creatures; although some places of the earth bee more inhabited then others.

But

of the Circles. 161

But as touching the latitude; if towarde the North in the country of Lapous, & the south (toward the vtmost coast of America shal end) seing ẏ vtmost distance of the earth hath very litle béene noted, of this shall small errour be caused.

If two places offered or giuen be placed vnder the Equatour, of which the space is sought, then the arke of the difference of latitude, is the same with the arke of the distance, neither doth the verticiall cy2cle differ from the Equatour. For the equatour of either place doeth containe the verticiall points, as may appeare in this tryangle, noted with A. B. C. Of which, if 15. germain miles be wrought into parts of the difference of longitude, and any scruples after remaine, deuide those by 4. For (by so many minutes of a degrée, doth a Germain mile answere) that the distance shall make. As Ptholomie writeth, of the places vnder the Equatour.

The high lande or mountaine of the Satyres, in the country of Syna, whose longitude is of 175 degrées, and no minutes, nor hathany latitude. Myrica an Ile of Ethiope vnder Aegipt, whose longitude is of 85. degrées, the angle of the difference of longitude betwéene the meridians of these places, is straight or right, and containeth a whole quarter or 60. degrées. The like are these places standing vnder the equatour. Colipolis a citty of India beyond the riuer Ganges, which hath the longitude 194. degrées, and 20. minutes, & Essina the greate Mart-towne of Aethiope vnder Aegipt, whose longitude is of 70. degrées, and 3. minutes. The angle of the difference of longitude (which the meridians of these compasse) is blunte, and containeth 94. degrées, and 17. minutes. Againe the same or the like meridians containe and make a sharpe angle of 43 degrées, as of the citty Nubarta of Taprobane

probane, which at this day is Sumatra, and Colipolis of Inde beyond or aboue Ganges: for it is distant from the west 122. degrees, and 20. minutes, and this containeth 164. degrees, and 20. minutes.

If two places be giuen, the one standing vnder the Equatour, and the other distant toward any other quarter from it. The first, that the angle of the difference of longitude is straight to these here placed. In that if two places giuen the one shall be vnder the equatoure, but the other distant from the same toward some quarter, then must the angle of the difference of longitude bee considered. If the same shal be right, then shal the distance of either place be the quadrant of the greatest cycle. As in this tryangle A.B.C. where the letter A. representeth the Pole of the equatour and the places giuen, that the one be standing in the point B. vnder the equatour, and the arke A.B. be the quadrant: and that the other consisteth in the letter C. the angle then of the difference of longitude, being C.A.B. is right. By Regio a montano de trangulis appeareth, that C.B. the ark of the distance of places which reacheth out right, is a quarter of the greatest cycle. Wherefore if the degrees bee multiplied by 15. and the minutes deuided by 4. the distance then shal be knowne. As for example, Nubarta of Taprobone hath the longitude 121. degrees, and 20. minutes, but no latitude: the city Pyse of the Tuscanes in Italie, hath the longitude 31. degrees, and 20. minutes almost, the latitude of 42. degrees, and 11. minutes: then the angle of the difference of longitude is right, for the difference is of 90. degrees, or a whole quadrant. These then multiplied by 15. do procreate or bring forth the distance to be of 1350. Germaine miles.

Essina a Mart-towne or principal citty of Aethiope vnder

of the Circles. 163

der the gouernment of Aegipt, hath the longitude of 70. degrées, and 3. minutes, but it hath no latitude. The Ile of Tyrus hath the longitude of 67. degrées, and no minutes the latitude of 33. degrées, and 20. minutes. The difference of longitude, betwéene the one and the other, is of 3. degrées, and 3. minutes. The complement of the difference of longitude, is of 86. degrées, and 57. minutes, of the latitude of the place not standing vnder the equatour, the complement is 56. degrées, and 40. minutes.

The royall citty Colipolis of Inde (aboue the riuer Ganges) hath the longitude of 164. degrées, and 20. minutes, but no latitude knowne. The longitude of Tyrus is of 67. degrées, and no minutes, the latitude hath 33 degrées, & 20. minutes. The difference of longitude greater then the quadrant, is of 97. degrées, and 20. minutes. The quadrant being abated, there remaineth 7. degrées, and 20. minutes. The complement of the latitude of Tyrus, is of 56. degrées, and 40. minutes. If of two places giuen, either standeth without the Equatoure toward some one of the opposite quarters; and the other vnder the equatoure: then is the reason of the standing considered, and the angle of the difference of longitude. For the one differeth either by like spaces from each bound, and is nearer to the Pole, the other to the Equatoure. The same appeareth by the compared latitudes, which like toppes of either place containe the same Parallel, the vnlike being distant, and the Parallell by a space seperated, toward each place, doe argue peculiar and proper tops. But the angle of the difference of longitude, either it is right, blunt, or sharpe. This of the placing and diuersitie of the angles, doeth much varie or alter the reason & methode of the searching of these.

If two places giuen haue equall arks of the latitudes, and from the middle or halfe of the equatoure bee alike distant, and how much so euer the angle of the difference of longitude be, as here vnder the difference of longitude is in the first, of the example taught: yet are the arkes of the latitudes agréeing and equally founde, so that in this example appeareth no difference, but in the only longitudes of the places offered. As for example.

The longitude of Danske is of 39. degrées, and twoe scruples or minutes: the latitude of the same hath 54. degrées, and 48. minutes. The longitude of Lubecke is of 28. degrées, and 20. scruples, the latitude hath 54. degrées, and 48. scruples. The difference of longitude consisteth, of 10. degrées, and 42. minutes. The halfe difference is, of 5. degrées, and 21. scruples. The distance on earth, betwéene Danske and Lubecke, is of 92 Germaine miles, and a halfe.

The great citie Alexandria vnder the Turke (after Ptolomie) hath the longitude of 122. degrées, the latitude of the same, is of 41. degrées. That famous Toletum or Toledo of Spaine, hath a longitude to the same, of 10. degrées, the latitude of the same is, of 41. degrées. The difference of longitude betwéene the one and the other, is of 102. degrées. The halfe difference, hath 51. degrées. The complement of the equall latitudes of either, is of 49. degrées. The whole distance betwéen both appeareth to containe 1077. Germaine miles and a halfe.

If of two places giuen, the one bee further distant from the equatour then the other, and the greatnesse of the complements of either latitude differing (as that the arkes of the latitudes be vnequall) so that the diuersity of the angle included with the arks of the complements, shal varie the methode or reason of the search, for that the one giueth and formeth a right angle, another a sharpe, another a blunt angle: yet to these, the angle of the difference of longitude is right. The example of two places differing alike (both in the longitude and latitude) here appeareth. The citty Tacola (which at this day is called Malchaia or Magna) a

place

of the Circles. 165

place of much resort of Marchants. This from the West hath the longitude of 160. degrees, and 30. minutes, of latitude from the equatour, it is 4. degrees, and 15. scruples distant. The other city and place in the countrey of Pontus (named Trapezus) being a head city of Cappadocia and was the aunctient seat of the Emperours. This hath the longitude of 70. degrees, and 30. minutes, and the latitude of the same is of 43. degrees, and 5. scruples. The difference of longitude betwéene the one and the other, is of 90. degrees. The arke of distance betwéene both places is of 87. degrees, and 6. minutes: to which 1306. & a halfe Germaine miles answere.

If the vnequall arkes of the latitudes, and angle of the defference of longitude be lesser then the right, it causeth a diuers reason of the search, by which the arke of the complement of the greater latitude doth varie thrée waies, as it is greater or lesser, and as with the arke by the second inquisition surely knowne, and beeing ioyned, forme either more or lesse a quarter of the cyrcle. Or thus, that the angle which the vnequall complements of the vnequall latitudes include, be sharpe; that is, and if the arks of the latitudes of either place be vnequall, and the difference of longitude bee lesser then the quadrant. As in this example more plainer appeareth, of twoe places beeing of sundry longitudes. That worthie citty Trapezus of Cappadocia, whose longitude is of 70. degrees, and 30. minutes, the latitude 43. degrees, and 5. minutes. The longitude of that well knowne city of Rome, hath 39. degrees, and 8. scruples, the latitude 41. degrees, and 8. minutes. The difference of longitude, betwéene the one and the other, is of 33. degrees, and 22. minutes. Another example not vnlike the former, and not much varying from the former: as the longitude of Ierusalem, which is of 66. degrees, and no minutes, the latitude, of 31. degrees, and 40. scruples. The longitude of Viteberge, being of 30. degrees, and 30.

M iij.   mi minutes, the latitude 51. degrees, and 50. scruples. The difference betwéene the one and the other of longitude is of 35. degrees, and 50 scruples.

If in places vnequally distant from the equatoure, the angle of the difference of longitude shalbe blunt, by which the difference of longitude shall appeare greater than the quadrant. Or thus, that the angle of the difference of longitude be blunt, séeing the places are further distant then a whole quarter, and thereby causeth a diuers reason and way of serch from the former; which semblably the diuers quantity of the complement of the greater latitude doeth thrée manner of waies varie, as in the same arke (which perfectly knowne by the second) is either greater or lesser. The example of this appeareth of these two places: the noble city Antiochia in Syria, which was after caled Seleucia, hath the longitude of 106. degrees, and no minutes, the latitude is of 40. degrees, and 40. scruples. The other of Toletum, whose longitude is of 7. degrees, and 4. scruples, the latitude hath 37. degrees, and 50. minutes. The difference of longitude is of 98. degrees, and 56. scruples, which deducted from the halfe cyrcle (or 180. degrees) the difference that remaineth vnto the halfe cyrcle, is of 81. degrees, and 4. minutes. The like example not much varying from the former of these two places: as the noble city of Portugale named Lysebone, whose longitude is of 4. degrees, and 18. scruples, the latitude hath 39. degrees, and 38. scruples. The other named Calecute (although the latitude differeth) hath the longitude of 112. degrees, and no minutes, the latitude is of 5. degrees, and no minutes. The difference of longitude, containeth 107. degrees, and 42. scruples more then the quadrant. The same deducted from the halfe cyrcle, doth expresse the difference remaining vnto the halfe cyrcle to bee of 72. degrees, and 18. minutes. The complement of the greater latitude, is of 50. degrees, and 22. scruples. The complement of the
lesser

lesser latitude, is of 85. degrées, and no scruples.

Another example of two places distant from the Equatour, of which the one is distant from the middle of the Equatour into the North, and the other into the South, as this example further instructeth; the one beeing the Ile of Thilen (which in Ptholomies time was the vttermost bond of the earth knowne Northward) that hath the longitude of 33. degrées, the latitude Northerly, of 63. degrées. The other called the Ile of S. Thomas, hath the longitude of 27 degrées, and 20. minutes, the latitude Southerly, of 16. degrées. The difference of longitude, is of 5. degrées, and 40. minutes. The complement of the latitude Northerly, is of 26. degrées.

A third example of the difference of other two places, as Basta of Taprobane, which Ptholomie affirmeth to bee in longitude 126. degrées, and in latitude toward the South 6. degrées, and 30. scruples. The other named Stocholma, in the Realme of Suecia, hath the longitude of 42. degrées, and 38. scruples, and the latitude of 60. degrées, & 30. scruples. The complement of the latitude Boreal, is 29. degrées, and 30. minutes.

The common way of measuring of places, with their spaces, by the rules of longitudes and latitudes.

Ere before I haue somewhat written of sundry habitable places on the earth, whose sundry points differ betwæne the one and the other; either in the onely longitude, or in the onely latitude, or in the longitude and latitude both together. Those places which do differ in the onely longitude, be distant by equal

M iiij. spaces

spaces from the equatoure, toward either of the Poles of the worlde: the verticiall pointes of those places ended by the same Parallell ioyning next the same space betwéene: yet each haue their owne proper meridians, being not distant by a like space from the Westerly bounde. The distance of these is alwaies gathered and noted in the same Parallel, which commonly belongeth to either place standing or hanging right ouer the tops of them.

Those places which doe differ in the onely latitude, are standing vnder the same meridiane, but they haue diuers Parallels, and each proper; and those continually distant vnequally, either towarde one pole from the middle of the Equatour (if either place declineth vnto one and the same quarter) or otherwise from the middle of the equatoure seuered and distant into the contrary quarters, by equall or vnequall spaces. If that one of the places looke into the South, and the other into the North, the distance of these is alwaies accompted in the common meridian.

Those places which do differ both in the longitude and latitude togither, or both decline towarde one Pole of the world, or seperated and distant from the midst of the equatoure towarde the opposite Poles (as the one looking into ye North, and the other into the south) or els by equal Parallels distant from the equatour; of which two onely are in the Sphere. If they bee reduced and applied vnto one great cyrcle (per 3. secundi Theodosij) or els bee vnder by vnequall Parallels, and by an vnequall space. The difference of the longitude of those (which either bee towarde them, or toward the Poles equally distant) is alway gathered in the middle Parallell betwéen either of the bonds by arithmeticall proportion, as afore taught. But in those places which haue equall Parallels, and equally dastant vnto the opposite quarters, the difference of longitude is imagined & noted in either of the equall Parallels. Therefore the arke hath the distance of the places standing, by the

the next space drawne ouerthwart by the pointes of those places, which with the arks of the differéce of either (both of the longitude and latitude) doth forme and make a sphericall tryangle right cornered, alwaies in the vpper face of the Globe. If that two meridianes meete and ende at the poles of the worlde, and beeing cut by the ouerthwart cyrcumferences of the Parallels, doe make with the included arkes of them right cornered tryangles, through the foure right lesser angles: but the angles beeing not right, the arke of the distance of the places doth deuide them into two right cornered tryangles. One of those tryangles is vsed in the common accompt for the right cornered; because in places not farre distant from the equatoure, the angles contained betwéene the mutuall sections of the meridians and Parallelles, doe not so much varie from the right angles: but in places far distant from the equatour, they varie very much. Now the rules for the diuers standing of places shall be taught in an easie and common maner.

If places doe differ in the onely longitude.

TO the searching and knowing of this, like as in the former, are the longitudes and latitudes of places giuen required; by which they being founde (seeing in the latitude there is no diuersity) the difference of longitude is onely to be considered, by deducting the lesser longitude out of the greater, and then howe many miles by proportion of the Parallell, vnder which the places stand or lie to the equatoure, answere to one degrée of the same. The same doth that rule (set forth in the fourme of a table here following) declare, beeing
drawne

drawne and made vnto this vse by the learned; in which, the miles that answere to one degrée of each Parallell, are there founde and noted vnto one degrée of the distance of the Parallell from the equatour. If to the whole degrées of the distance of the Parallels doe minutes depend, then from the difference of the two next numbers to one degrée, may the proportionall part be deducted or drawne: which from the number of the miles expressed vnto a whole degrée is abated, that the Parallels succéeding, may by litle & litle be caused to streach & appeare narower. To be briefe, the miles with the scruples or quarters (if any bee adioyned) let them bee reduced into the whole arke of the difference of longitude, which then shall manifestly shewe and expresse, the measured space by the Germaine miles.

Ptholomie when he had learned the longitudes and latitudes of certaine notable places, he could extract and gather by them the other vnknowne places, by the distances truly learned from trauailers. For by the longitudes and latitudes knowne of two cyrcles, and the distance also of them from any third place, there is then offered and giuen to know, as well the longitude, as the latitude of the third place. Further, in any two places lying and being in the vpper face of the earth, are fiue notes commonly learned. The distance of them, conuerted into degrées: the latitude of the one, and the latitude of the other : the difference of longitudes: the angle vnder the circumferencial distance: and the meridians contained by the other. Of the which fiue, if thrée onely be knowne, it is certaine that the other two may easily come to knowledge by the practise and skil of the sphericall tryangles.

An example of these former wordes (as touching the difference of longitude of two places) the latitudes beeing alike. As the city Byzantium nowe called Constantinople, whose longitude is 55. degrées, and no minutes, the latitude hath 43. degrées, and 5. minutes. The other city Trapezus

of the Circles. 171

pezus hath the longitude of 70. degrees, and 50. minutes, the latitude of 43. degrees, and 5. minutes. The difference of longitude is of 15. degrees, and 50. minutes to one degree of the common Parallell, and to each place, doe 10. Germaine miles with ⅔. answere or agree. These now brought into the difference of longitude, doe cause & make 174. Germaine miles almost. The like example to the former, is Arbela of Assiria, which hath the longitude of 80 and no minutes, the latitude of 37. and 15. minutes. The other Athens, whose longitude is of 52. degrees, and 15. minutes, the latitude 37. degrees, and 15. minutes. The difference is of 27. degrees, and 45. minutes.

Other briefe examples.

	Lon.	Lati.
Areca in Comagena being a part of Syria.	70. 10.	37. 15.
Megara the country of Euclide.	52. 0.	37. 15.

The difference is of 18. 10.

	Lon.	Lat.
Philippi a city in Thracia or country of Alexandria.	50. 45.	41. 50.
The royall city of Roome.	36. 20.	41. 50.

The difference is of 14. 25. longitude.

Lipsia,	29. 58.	51. 24.	The difference of lõg.
Antwarpe,	20. 16.	51. 28.	is 9. degr. & 42. min.
Vratislauia,	34. 34	51. 10.	The difference of lõg.
Erphordia,	28. 30.	51. 10.	is 6. degrees, & 4. min.

If places doe differ in the onely latitude, or that both be placed toward one pole, or either distant from the middle of the equatour, so that in the oncle latitude the places differ, when the longitudes be like, the standing of the places is to bee considered towarde either Pole, whether either place declineth toward one Pole, or that the one be Southerly, and the other Northerly. If they decline vnto one place and quarter, then deduct the lesser latitude, out of the more, and the difference of latitude shall appeare. If eyther be distant from the middle of the equatoure, the latitudes ioyned doe shew the difference. The degrées, of the difference wrought by 15. and the scruples deuided by 4. shall offer & giue the estimate distance in Germaine miles. As in this example, the city of Noriberge hath 28. degrées, and 20. minutes of lōgitude, the latitude is of 49. degrées, and 24. minutes. The other is Mylayne, whose longitude is of 28. degrées, and 20. minutes, the latitude hath 45. degrées, and 6. minutes. The difference of latitude, is of 4. degrées, and 18. minutes: the space betwéene, is 64. miles and a halfe. Like examples are these.

Trapezus,	70. 15	43. 5.	The difference of latit.
Antioch,	70. 15.	37. 20.	is 5. degrées, & 45. min.
Padua,	31. 50.	51. 0.	The difference of latit.
Budissina,	31. 50.	44. 16.	is 6. degrées, & 44. min.

If two differ together in the longitude and latitude, and that either declineth towarde one Pole, then in either toward the places differing, as in the longitude and latitude are the differences of the spaces from either bounde of the latitude and longitude gathered. The halfe difference of the latitude added to the lesser altitude shall shew the Parallell in which the difference of longitude is accompted. With that Parallell by this rule are the miles gathered and knowne, which answere or agrée to one degrée. These founde, reduce into the whole difference of the longitude, and that which procédeth (agréeably) of the same; that is, multiplied in it selfe or arising of the multiplication kéepe. After the degrées of the difference of longitude reduce into 15. and the minutes annexed (if any such be) distribute or

diuide

of the Circles. 173

deuide by foure, that which ariseth of either working, reduce ioyntly one to the other, and adde to the number kept afore. For of the whole gathered may the square roote be attained, which sheweth the distaunce of places. As by a like example, the city of Witeberge hath the longitude of 30. degrees, and 30. minutes: the latitude of 51. degrees, and 50. minutes. The other being Ierusalem hath 66 degrees of longitude, and no minutes: the latitude is of 31. degrees, and 55. minutes. The difference of longitude is of 35. degrees, and 30. minutes: the difference of latitude is of 19. degrees, and 55. minutes. The middle Parallel in which the difference of longitude is accompted, doth differ or is distant from the equatour 41. degrees, and 52. minutes: to one degree of the same doe 11. miles, and 10. scruples of a Germaine mile answere, which reduced into the difference of longitude, doe procreate or bring foorth 396. Germaine miles: these wrought together make 156816. The degrees of the difference of latitude being wrought by 15. & the scruples deuided by foure, doe make 266. Germaine miles; which multiplied one in the other, do performe and make 89401. Either of those square numbers ioyned, and the roote extracted, the distance shall appeare to be 495. miles.

The finding of the distances of places
or citties, in a more easier maner.

THat you may knowe howe by the longitudes and latitudes of twoo places or citties, the distaunce of them may be found: thus do, when two cities be offered (whose largenesse is to you vnknowne) seek the longitude and latitude of both by the Cosmographie of Apian, or Ptolomies

lomies Geography; which being found, write downe the longitude of the one vnder the longitude of the other, and the latitude of the one vnder the latitude of the other (as the former examples shew) in such sort, that the degrees of the other, and likewise the minutes vnder the minutes. After seeke the difference as well of the longitudes and latitudes, in this maner: subtract the lesser longitude from the greater, the remainer is called the difference of the longitudes. After deduct the lesser latitude out of the more, and the difference of the latitudes shall remaine. By the differences of the longitudes and latitudes, shall the distance of cities giuen be gathered. But in that there is three maner difference of places, as that there be certaine places which differ in the onely latitude; that is, vnder one meridiane, and yet lie vnder diuers Parallels: and certaine that differ in the onely longitude; that is, vnder one Parallell: yet are diuers meridians: and certaine that do differ both in the longitude and latitude; that is, they lie vnder diuers meridians, and Parallels, three rules also of the searching of distances, betweene two places, are taught of the Geographers.

The first rule.

When two cities hauing one longitude are offered (but hauing sundry latitudes) deduce the lesser out of the more: the rest of degrees, in that they be the degrees of the great cycle, multiply by 15. (for that 15. Germaine miles answere to one degree of the great cycle) and then shall you haue the distance of the cities.

But if minutes depend to the degrees of difference, then deuide them by foure, the quotient adde to the fore number of the miles. For seeing one degree or 60. minutes do make 15. Germaine miles; it ensueth, that foure minutes make

of the Circles. 175

make one Germaine mile, &c.

An Example.

MAdeburge and Egra agrée only in longitude; that is, they bée equally distant from the West or from the meridian, which is drawne or stretched by the fortunate Iles. For the longitude of either towne is of 29. degrées, the latitude of Madeburge is of 52. and 20. minutes: the latitude of Egra is of 50. degrées, and 5. minutes: therefore is Egra more Southerly then Madeburge. The difference of the latitudes, is 2. degrées, & 15. minutes; that is, 33. Germaine miles, with a halfe & a quarter of a Germain mile.

Another.

THe longitude of Trydent is of 30. degrées, and 30. minutes. The longitude of Viteberge is asmuch. The latitude of Trydent is of 45. degrées, & 14. minutes. The latitude of Viteberge is of 51. degrées, and 50. minutes. These now differ in the onely latitude, which difference of the latitude is of 6. degrées, and 36. minutes; that is, 99. Germaine miles. So much is the distance almost betwéen Trydent and Viteberge.

Another.

THe longitude of Thunis is of 36. degrées, and 50. minutes: the longitude of Salerne in a maner the same. The latitude of Thunis is of 32. degrées, and 30. minutes. The latitude of Salerne is of 40. degrées, and 30. minutes.

The

The difference of latitude is of 8. degrées, & no minutes, that is, 120. miles. And somuch is the distance betwéene Thunis and Salerne,

Another.

The City of Yorke, and the Towne of Barwicke, agrée in longitude: for the longitude of either place, is of 17. degrées, and no minutes. But they differ in latitude, in that the latitude of Yorke is of 54. degrées, & no minutes, the latitude of Barwicke, is of 56. degrées, &50. minutes. The difference of the latitude is of 2. degrées, and 50. minutes: that is, 210. English miles. So much in a manner is the distaunce, betwéene the City of Yorke, and Barwicke.

Another.

The City of London and Northampton, in a maner is of like longitude. For the longitude of London is of 16. degrées, and 30. minutes approued. But they differ in latitude, in that London hath the latitude of 51. degrées and 34. minutes, the latitude of Northampton is of 52. degrées, and 50. minutes. The difference of the latitude, is of 1. degrée, and 16. minutes; that is, 76. English miles. So much in a maner is the distance betwéene London and Northampton.

Another.

This example differeth both in the longitude and latitude somewhat. For the longitude of Colchester, is 18. degrées, and 30. minutes, the longitude of Oxeforde hath 15. degrées, and nominutes. The difference of longitude betwéene the one and the other, is of 3. degrees, 38 minutes, that is, 109. English miles. The latitude of Colchester hath 51. degrees, and 59. minutes. The difference

rence of latitude, is no degrees, and 16. minutes. So that 16. English miles, is the distaunce betweene the one and the other, after their standing Northward.

Another.

Cygnea and Ratisbone, agree in longitude, for either is of 29. degrees, and 51. minutes: but they differ in latitude, in that the latitude of Cygnea hath 50. degrees, and 46. minutes, the latitude of Ratisbone, of 48. degrees, and 56. minutes. The difference of latitude betweene the one and the other, is 1. degree, and 50. minutes, which make 27. and a halfe Germaine miles.

The second rule.

Before the second rule be here taught, it behoueth that you know howe many Germaine miles aunswere to each degree of the parallel (passing by the Zenith of Cities offered.) Here conceiue that not as in the former rule, to euery degree of each parallell, but to each degrees onely of the parallell Cyrcle, which streacheth and is vnder the Equinoctiall, and as principall of all the parallels, deuideth the whole earth into twoe equall halues, to which are 15. Germaine miles attributed, as to a degree of it. Where the other cyrcles (as afore written) be not of the same bignesse, but how much nearer they be to the poles, so much the lesser they are: and how furder of they be frō the poles, so much the greater they are. Whereof it is manifest, that aswell the greater as the lesse Cyrcle of the parallels, is distributed or deuided into 360. degrees, and that those degrees (according to the distance of those parallels from the poles) be greater or lesser.

For the same cause shall you here finde in the table following, how many Germaine miles answere in each eleuations, to the degrees of the parallels.

N i.   A

The second Part

A Table, containing the degrees of the differences of each Paralels, from the Equator vnto the proper Pole, by whole degrees of the Latitudes conuerted into Myles.

Degrees.	Myles.	Scruples.	Degrees.	Myles.	Scruples.	Degrees.	Myles.	Scruples.	Degrees.	Myles.	Scruples.	Degrees.	Myles.	Scruples.
1	14	59	19	14	11	37	11	59	55	8	30	73	4	23
2	14	59	20	14	6	38	11	49	56	8	23	74	4	8
3	14	58	21	14	0	39	11	39	57	8	10	75	3	53
4	14	58	22	13	54	40	11	29	58	7	57	76	3	38
5	14	56	23	13	48	41	11	19	59	7	43	77	3	22
6	14	55	24	13	42	42	11	9	60	7	30	78	3	7
7	14	53	25	13	36	43	10	58	61	7	16	79	2	52
8	14	51	26	13	29	44	10	47	62	7	2	80	2	36
9	14	48	27	13	22	45	10	36	63	6	48	81	2	21
10	14	46	28	13	15	46	10	25	64	6	34	82	2	5
11	14	43	29	13	7	47	10	14	65	6	20	83	1	50
12	14	40	30	12	59	48	10	2	66	6	6	84	1	34
13	14	37	31	12	51	49	9	50	67	5	52	85	1	18
14	14	33	32	12	43	50	9	38	68	5	37	86	1	3
15	14	29	33	12	35	51	9	26	69	5	23	87	0	47
16	14	25	34	12	26	52	9	14	70	5	8	88	0	31
17	14	21	35	12	17	53	9	12	71	4	53	89	0	16
18	14	26	36	12	8	54	8	49	72	4	38	90	0	0

of the Circles. 279

An Example for the vſe of this Table.

Vneburgum and Stetinum, haue the eleuation of the Pole precisely of 54. degrées, to knowe howe many Germaine miles aunswere to one degrée of the Parallell, passing by the Zenith of either Citty, enter your Table, and there diligently loking, you shall finde by the degrée of that latitude 54. noted eight miles, and 49. scruples of a mile. For so many miles in that Parallell answere to a degrée; that is, eight, a halfe, and the third parte almost of a Germaine mile. And this is easily found, if the eleuation doth onely consist in whole degrées. For in each eleuation are certaine miles, and the scruples of a mile, answering to each degrée assigned. But if the place or city haue minutes depending to the latitude as Viteberge whose lattitude is of 51. degrées, and 50. minutes: then séeke in this table how many miles and scruples of a mile, are atributed to the whole degrées, and you shal finde by the degrée of the latitude of 51. noted 9. miles and 26. scruples of a Germaine mile. After séeke the miles and minutes that nexte ioyne to the eleuation following, being 52. and you shal find right against 9. miles, and 14. scruples of a mile: which so set down or placed, ỹ the miles bee vnder the miles, and the minutes vnder the minutes, after this maner.

```
           miles    minutes
             9.       26.
             9.       14.
```

Subtract the lesser number out of the more and vpper written, and there will remaine 12. minutes, of this rest; that is, of the 12. minutes, séeke the number proportional, according to the proportion of one degrée or 60. minutes, vnto the minutes depending to the latitude offered, as of the

the latitude of Viteberge to the whole degrees, do 50. scruples depend. Of which so place the numbers by the Rule of three, working and saying on this wise, 12 Z 50. that if one degree or 60. minutes of the degree 60 — 10. doe giue 28. minutes of a mile, how many scruples of a Germaine mile, doe 12. minutes of a degree giue. To know this, multiply the first by the second, that is, 12. by 50. & the increase shalbe 600. this product diuide by the thirde number which is 60. and the part proportional shalbe 10. This proportional part found, subtract out of the miles, and minutes of the former eleuation; that is, from the 9. miles, and 26. minutes: deduct the 10. and there will remaine 9. miles, and 16. scruples, precisely answering to one degree of the Parallell passing by Viteberge. Here the second rule followeth, which is easie to conceiue, if you worke according to the former taught.

The second Rule.

If two Citties be offered, which differ in the only longitude, first seeke by the instruction aboue taught, ye miles, and minutes of a mile answering to one degree of the Parallel, passing by the Zenith of those Citties. After, seeke the difference of longitudes in the degrees and minutes: then multiply the difference of longitudes, with ye miles and scruples of the miles, and the distance shall appeare of the Cities giuen.

An Example.

Viteberge and Westphalia agree in latitude: that is, they be both standing vnder one Parallell. For the latitude of Viteberge is 51. degrees, and 50. scruples, and exceedeth the latitude of Westphalia by certaine minutes, which here we passe, but they differ in longitude, in that Westphalia lies more to the West. The longitude of Viteberge is 30. degrees, and 30. scruples: the longitudes of Westphalia is 24. deg. & no min. To find the distance, see how many miles answere to one degree of longitude in ye parallel, passing by the Zenith of the Citties giuen. Be-

of the Circles.

fore was taught, that in the Parallel of Viteberge 9. miles and 16. scruples do answere to one degree: wherefore seeke the difference of longitudes of the two Cities, and deduct the lesser number out of the more; that is, let the 24. degrees and no minutes bee deducted from the 30. degrees, & 30. scruples, & the difference resting, shall be of 6. degrees, and 30. scruples. Last, multiply the 6. miles and 16. scruples, with the difference of longitude; that is, with 6. degrees, and 30. minutes, and you shall haue the distance of the twoe Cities. But here obserue and note diligently in the multiplicatiō of the degrees, miles, and minutes, what procedeth and commeth of the same. For the miles multiplied by the degrees, doe bring foorth the miles: and the miles multiplied by the minutes of the degrees, doe bring forth the scruples of the miles. The minutes of the miles multiplied by the degrees, doe produce or bring foorth the minutes of the miles. And last, the minutes of the miles multiplied by the minutes of the degrees, doe produce the seconds of the miles.

But that this may the readier be conceiued, vse this example, the former Westphalia and Viteberge: where the 9 miles and 16. scruples, are to bee multiplied by the 6. degrees, & 30. minutes on this wise. Multiply the 9. whole miles, by the 6. whole degrees, thus: as sixe time 6. bringeth out 54. miles. Multiply after that, the whole miles by the minutes of the degrees; thus, that 9. times 30. doe make 270. minutes of miles. After multiply the minutes of the miles by the whole degrees, and by the minutes of the degrees: as the 16. minutes of the miles multiplied by the 6. degrees, doe make 99. minutes of miles. After this the 16. minutes of the miles multiplied by 30. minutes of the degrees, doe make 480. secondes of miles; which minutes and seconds gather into whole miles, in this maner. First deuide the 480. seconds by 60 and the quotient shall be 8. minutes. (For that one minute containeth 60. seconds, as one degrée doth cōtaine 60. minutes. These 8. minutes, adde to the minutes procéeded of the former or vpper working; that is, the 270. and the 96. ¶ you shall haue 374. scruples of miles, which deuided by 60. the quotient will be 6. whole miles, and 14. scruples; that is, almost the fourth part of a Germaine mile. These miles gathered of the seconds and minutes of the miles, adde to the 54. miles gathered afore by the multiplication of the degrées and miles, and you shall haue the true distaunce betwéene Viteberge and the Monasterie of Vestphalia; that is, 60. Germaine miles, and almost a quarter.

This maner of working in searching the distance of places (which differ in the onely longitude) obserue in the other examples following: in which you shal finde their distance, by hauing their longitudes and latitudes.

Here folowing shall be sundrie examples, in which the young students and practisers may excercise them according to rule.

An Example.

COleine and Marburge do differ in the only longitude: for the longitude of Coleyne is of 23. degrées, and 28. scruples, the longitude of Margburge hath 25. degrées, and 45. minutes. The latitude of either (which agrée) is of 51. degrées, and no minutes. The difference of longitudes is of 2. degrées, and 17. minutes. The miles answering to one degrée (drawn in that Parallell by the Zenith of the Cities giuen) are 9. miles, and 26. scruples, as may appeare in the former table. But séeing no minutes depend to the latitude, the 9. miles, and 26. minutes are to bee multiplied by the difference of the longitudes: that is, the 2. degrées, and 17. minutes, in this manner: saying

ing twice 9. doe make 18. miles, twice 29. are 52. minutes of miles, nine times 28. doe make 152. minutes of miles, and seauentéene times 26. are 442. secondes of miles: which secondes and minutes deuided by 60. doe make thrée miles, 32. minutes and 22. seconds. These added vnto the 18. miles, declare the distance of Coleyne and Margburge, to bee of 21. Germaine miles, and a halfe.

Another.

The longitude of Franckeforde is of 25. degrées, and 38. minutes. Hasforde is of longitude 37. degrées, and 52. scruples. The latitude of either, is of 50. degrees, and 12. minutes. Nowe they differ in the onely longitude, for that the difference of the longitudes is, of 2. degrées, and 14. scruples; that is, Franckeforde by twoe degrées, and 14. minutes, is more towarde the West, than Hasforde. The miles according to latitude 50. are 9. and 38. minutes, and the miles according to the latitude following, as 51. are 9. and 26. minutes. The difference of these twoe manner of miles and minutes, is 12. minutes: the parte proportionall subtracted, is twoe. The miles answering to one degree, in the Parallell drawne by the Zenith of Franckeforde and Hasphorde, are 9. and 36. minutes. Nowe as aboue these miles and minutes (with the difference of the longitude) that is, twoe degrées, and fouretéene minutes multiplied, you shal haue the distance in Germaine miles; that is, twenty and two, and almost a halfe.

Another.

The longitude of Gawnt (the natiue towne of Charles the

the first Emperour) of 19. degrées, and 8. minutes. The longitude of Lipsia of 29. degrées, and 28. minutes. The latitude of either is of 51. degrées, and 24. minutes. The difference of longitudes is of 10. degrées, and 50. minutes. The miles according to the eleuation 51. are 9. and 26, minutes: the miles ensuing the eleuation assigned, are 9. and 14. minutes. The difference of these two manner of miles and minutes, is 12 minutes: the part proportional subtracted, is 4. minutes. The miles answering to one degrée in the Parallell (to Gaunt or Lipsia) are 9. and 22. minutes. These miles and minutes multiplied with the difference of the longitudes, do offer and shew the distance betwéen Gawnt and Lipsia; that is, 101. Germaine miles, and almost a halfe.

Another.

The longitude of Straseborow is of 24. degrées, and 30. minutes, the longitude of Landunum of Bauier is of 30. degrées, and 25. minutes. The latitude of either is of 48. degrées, and 45. minutes. The difference of longitude, is 5. degrées, and 55. minutes, &c.

Another.

The longitude of Direpsa is of 130. degrées, and no minutes; the longitude of Danaba of 104. degrées, and no minutes neither. The latitude of either is of 45. degrées, and no minutes. The difference of longitude, is 26. degrées, and no minutes.

An easier working.

If this curiosity in obseruing minutes trouble you, you may then with lesser paines and errour leaue them, especially

of the Circles. 185

cially in places beeing not far distant a sunder, where the minutes omitted doe litle force or hinder, howe neare soeuer you finde the true distance. And by this meanes the second rule, is of no difficulty: for that euery painefull labor doth especially consist in the multiplying of the difference of longitudes, with the whole miles offered by the former Table, according to the degrée of latitude, of the Cities giuen.

An Example.

AMsterdame and Brandenburge (which as vnto whole degrées appartaineth) agree in latitude: for the latitude of either place in whole degrées, is 52. degrees. But they differ in longitude, in that the longitude of Amsterdame is 21. degrees, and 4. minutes, the longitude of Bradenburge of 30. degrees, and 35. minutes. They differ in longitude 9. degrees; that is, Amsterdame is nearer to the West then Bradenburge by 9. degrees, as the former table teacheth in the Parallell of the latitude 52. which containeth 9. miles. Now by so many miles is Bradenburge distant from Amsterdame.

Another.

NOrdlinga and Nicostadium, agree in latitude, for the eleuation of the Pole, or latitude of either is of 48. degrees. But they agree not in longitude, in that the longitude of Nordlinga is of 27. degrees, and 54. minutes, the longitude of Nicostadium of 29. degrees, & 32. minutes: so that they differ 2. degrees, which make 20. Germaine miles, as may appeare by the fourmer table, where 10. miles are assigned to the latitude 48. Now you shall vnderstand that the distance of Nordlinga and Nicostadium, is of 20. Germaine miles almost.

An-

Another.

The longitude of the City of Venice is of 32. degrées, and 30. minutes, the longitude, of Spoletum is of 36. degrées, and 30. minutes. The latitude of either, is of 44. degrées. The difference of the longitude is of 4. degrées. And 10. miles doe answere to one degrée in the Parallel of the latitude 44. The miles being multiplied by the difference of the longitudes; that is, by foure degrées, doe declare the distaunce of Venice and Spoletum, to bee of forty miles.

If of two places, the one being Sou-
therly, and the other Nor-
therly.

If of twoe places giuen, the one hath a latitude Northerly, and the other a latitude Southerly: séek the difference of either space of the longitude; after subtract the lesser longitude out of the greater (but of the latitude Northerly and Southerly) according to the latitudes ioyned of either place. In the second place the standing must bee cosidered, whether they be scituated vnder equall Parallelles, and both distaunt by a like space from the Equatoure, or else otherwise seperated by vnequall Parallelles, and by an vnlike space. For if the Paralles of the places giuen shall bee equall, then must the difference of longitude be accompted in either alike: but if vnequall (and that both shall bee distant by an vnlike space) then the halfe of the greater latitude applied to the lesser latitude, shall demonstrate and shewe
the

of the Circles.   187

the Parallell apte and méete to this instruction: with the same Parallell are the degrées answering to each degrée, declared by the former rule, and the other is taught & shewed, as in the precedent place is declared.

Meroe a Region of Aethiopia vnder Aegypt, hath the longitude of 91. degrées, and 30. minutes, the latitude of 16. degrées Northerly.

The Ile of S. Thomas in the bordure of Aphrica hath the longitude of 27. degrées, and 20. minutes, the latitude Southerly is 16. degrées. The difference of longitude is 34. degrées, and 10. scruples. The difference of latitude which the conioyned latitudes do make) is of 32. degrées. And séeing both by an equall space bée distant toward the opposite poles from the equatour, it therefore forceth not, that the difference of longitude bee gathered in either Parallell Northerly or Southerly, in that they be equall. For to one degrée of the Parallell (which is of 16. degrées, distant from the equatour) doe 14. miles, and 25. scruples, answere or agrée: which reduced into the difference of longitude, doe bring forth 492. Germaine miles; which multiplied togither, doe bring forth 241064. The difference of latitude wrought or multiplied by 15. doe bring foorth 480. Germaine miles, which againe wrought togither do cause 230400. And by either quadrant conioyned, the square roote drawne out of the same, doth then declare and shew the distance to be of 686. Germaine miles.

The Ile of Thylen hath the longitude of 33. degrées, the latitude is of 63. degrées Northerly. The Ile of S. Thomas hath the longitude of 27. degrées, and 20. scruples, the latitude Southerly, of 16. degrées. The difference of longitude, hath 5. degrées, and 40. scruples, the difference of latitude, is of 79. degrées.

The halfe difference of the greater latitude, applied to the Southerly latitude, bringeth foorth that the Parallell is distaunt from the equatoure 47. degrées, and 30.
scru-

scruples: in which the difference of longitude, must be accompted. And to one degrée of it in the rule, doe 10. Germaine miles, and 7. scruples answere; which wrought into the difference of longitude, do bring forth 57. Germain miles almost. Those multiplied, doe make the increase to be 3239. By the difference also of the latitude, those multiplied, doe bring forth 1404225. Of the quadrants ioyned, the roote hath 1189. Germaine miles almost, that is, the distances sought of the places.

That the studious and diligent practisioners may easier perceiue and perfecter vnderstand these differences of the standing of places, let them often accustome thēselues therein, that when the longitudes and latitudes of sundry places be offered, they then consider whether they differ in the onely longitude, or latitude only, or in both, and what the latitude is of either, and into which parte from the Equatour: and besides that, they learne to expresse the standing of them by proper lines drawne, and the places noted.

If the numbers of the latitudes be alike, and the numbers of the longitudes be vnlike, then doe the places onely differ in the longitude. Therfore by two meridians found and defined, lying crosse to them in one Parallel, imagine and set the place of the greater longitude in the pointe of the crossing further off, that the other in the nearer may be placed vnto the West. For the place alwaies (whose longitude is lesser then the other) is nearer founde to the West, and the other is further distant into the East. The arke also of the Parallell included betwéene either meridian, doth demonstrate the difference of longitude.

If the numbers of the longitudes shall be alike, and the numbers of the latitudes vnlike, then is the diuersity of the places in the onely latitude. Therefore two Parallels drawn crosse, of which the one being higher and the other lower and crossing them by one meridian, they doe set the place

place of the greater latitude in the uppermost point of the crossing, and the other in the lowest point.

If the latitudes be alike, as the one Southerly, and the other Northerly: then the middle arcke of the Meridian being betwéene, is equally crossed by the Parallels of the places drawne thwartly by the Arke of the Equatour, in such sort, that the Equatour is by an equall space distant from either.

If both the numbers of the longitudes and latitudes shall be unequall, and either place distant into the North from the Equatour, therefore in both is there a diuersity. Therefore two Meridians being imagined, the one Orientall dextre, and the other Occidentall synistre, and that by so many Parallels drawne thwartly, which crosse the Meridians, the one Southerly, the other Northerly: and that the place whose greater longitude is touched in the lowest and furthest point, and the other to be noted right against; that is, in the upper and neerest point. Or thus contrariwise: If one place shall exceed the other, both in longitude and latitude, and be further standing in the higher pointe of the crossing, and thereby more farther distant, and the other noted to stand right against, and the seates also of the places unequally touch, which declareth and containeth the nighest distance of such places. In the same maner is the standing of places descending unto diuers partes from the Equatour expressed; being obserued in such order, that if the places of either be alike distaunte from the Equatour, the Equatour then is exquisitly standing in the middle of both: but if the places happen to bée unequall, then is the Equatour by an unequall distance, placed farther off.

A.

A third rule.

If twoe Citties offered doe differ both in the longitude and latitude, seeke first the difference aswell of the longitude, as latitude. After halfe of the difference of latitudes adde vnto the lesser latitude, and with the produce enter the table which in the former examples hath bene taught and practised: searching there the miles and minutes aunswering properly to one degree. The miles and minutes found, multiply with the degrees of the difference of longitude, and the produce multiply in it selfe, and you shall obtaine and haue the first quadrate. Thirdly, multiply the difference of latitude by the 15. Germaine miles, and this produce also multiply in it selfe, and you shall haue the second quadrate. Last, ioyne or adde togither these two quadrate numbers (and of that produced or encreased) search out the quadrate roote. The quadrate or square roote, is the distance of Cities offered.

An Example of the third rule.

Vuschegarda and Verona, do differ both in the longitude and latitude, in that the longitude of Vuischegarda is of 41. degrees, and 17. minutes, the latitude is of 52. degrees, and 4. minutes. The longitude of Verona hath 31. degrees, & 18. minutes, the latitude is of 44. degrees, and 49. minutes. The difference of the longitudes is of 9. degrees and 59. minutes. The difference of the latitudes is of 7. degrees and 15. minutes. The halfe of the difference of the latitudes, is 3. degrees, & 37. minutes, which halfe added to the lesser latitude; that is, to Verona, which is of 44. degrees, add 49. minutes, doeth then produce or bring forth 48. degrees, & 26. minutes. This produce or increase

is

of the Circles. 291

is named the middle latitude, in that it is distant by equal degrees and minutes from either latitude of Vuischegarda and Verona; that is, it exceedeth the latitude of Verona by 3. degrees, and 37. minutes, and Verona doeth exceede Vuischegarda by so many degrees, and minutes. With this product or middle latitude; that is, with 48. degrees, & 26. minutes. I enter the former table, and according to the instruction afore taught in the second rule, I finde in the parallell which is drawne by the middle latitude, to answer to one degree right against 10. Germain miles, and 2. minutes. It was also taught in the second rule, ỹ if minutes depended to the latitude, that those should be sought in the former table, and by the next eleuation folowing, the proportionall part to be sought. As in this example. The latitude 52. are 9. miles, and 14. scruples noted, and in that 3. degrees, and 37. scruples depende to a middle latitude, I seeke in the table how many miles and scruples are noted next to the latitude folowing, 55. and there I finde 8. miles, and 36. scruples. The difference betwene the miles and scruples of the eleuations of 52. and 55. is 1. degree, & 22. minutes. By the proportion of this difference, is the proportionall part gathered and founde, according to the maner afore taught in the second rule.

Another example of this third rule for thy further instructing of Viteberge and Lipsia, which differ in the longitude and latitude: for the longitude of Viteberge is of 30. degrees, and 30. minutes, the latitude hath 51. degrees, and 50. minutes. The longitude of Lipsia is of 29. degrees, and 58. minutes, the latitude hath 51. degrees, and 24. minutes. The difference of the longitudes is of thirty twoe minutes, the difference of the latitudes is of twenty sixe minutes. The halfe of the difference of the latitudes is of thirtéene minutes, which halfe added to the lesser latitude (as to Lipsia) which is of 51. degrees, and 24. minutes, doth produce 51. degrees, and 37. minutes.

The

The product is caled the middle latitude, in that by equal minuts it is distant from either latitude of Viteberge and Lipsia; that is, it excéedeth the latitude of Lipsia, 13. minutes, and by so many minuts is it excéeded of Viteberge. With this product or middle latitude; that is 51. degrées and 37. minutes, I enter the former Table, and by the Instruction afore vttered in the second rule, I finde in the Parallel which is drawne by the middle latitude, that 9. miles, and 19. scruples doe answere there to one degrée. And in the second rule afore is taught, that if minutes depende to the latitude, which is sought in the former Table, then by the next eleuation must the part proportional be sought. As in this example to the latitude, 51. degrées, are 9. miles and 26. scruples noted. And in that 37. minutes depend to the middle latitude, I therefore séeke in the table how many miles and scruples are assigned to the latitude next following; that is, 52. degrées: right against which I finde noted 9. miles, and 14. scruples. The difference betwéene the miles and scruples of the eleuations of 51. and 52. is of 12. minutes: so that by the proportion of this differéce vnto the whole degrée, or 60. minutes, is the proportionall parte drawne or gathered, according to the manner afore taught in the second rule. As thus, that as 69. minutes yéelde 12, euen so doe 37. giue 7. minutes, which is the parte proportionall. The same minutes subtracted from the miles and scruples assigned to the latitude 51. that is, from the 9. miles, and 26. scruples, there remaine 9. miles, and 19. scruples. And so many miles and scruples in the Parallel of the middle latitude doth answere vnto one degrée. Which being founde and knowne, these nine miles and the scruples, with the differences of longitude, which is of thirty two minutes, I then multiply, and they shew and bring forth 298. minutes: which multiplied againe in it selfe, do bring forth the first quadrate to be 88804. minutes. And this is ye first part

of the Circles. 193

parte of the working of these. Nowe followeth the other part.

I multiply first the difference of latitude, as the 26. minutes by 15. Germaine miles, and they bring forth 390. minutes, which multiplied againe in it selfe doe yælde 151200. minutes, as the second quadrant is. Now these two numbers quadrate added, doe bring forth and make 240904. minutes, of which the quadrate or square roote is of 494. minutes of miles. These for that they are the minutes of miles, ought to be deuided by 60. and then they bring forth 8. whole miles, and 14. scruples; that is, a fourth part almost of a Germaine mile. So that somuch is the distance, betwéene Viteberge and Lipsia.

Another.

The longitude of Buda is of 37. degrées, and 44. minutes, the latitude hath 47. degrées, and no minutes. The longitude of Aquisgranum is of 22. degrées, and 24. minutes, the latitude hath 51. degrées, and 6. minutes. The difference of the longitudes, is of 15. degrées, and 20. minutes. The difference of the latitudes is of 4. degrées, and 6. minutes. The halfe of the difference of the latitudes, is of 2. degrées, and 3. minutes. The middle latitude is of the degrées, and thrée minutes: here (in that 3. minutes doe onely depend to the middle latitude) are omitted, séeing the leauing of them bring or cause small error. Then must you take the miles assigned to the latitude 49. that are 9. miles, and 50. scruples, which with the difference of the longitude; that is, 15. degrées, and 20. minutes are to be multiplied, and they shall bring forth 150. miles, and 46. scruples: which miles containe as a quadrate; that is, one parte in it selfe with the minutes, that may bee multiplied and resolued also into minutes in the multiplication by 60. it shall then bring forth 9000. minutes.

nutes to these adde the 46. minutes, and the number then shall be of 9046. minutes. These minutes againe multiplied in it selfe doe bring forth and offer the first quadrate, that is 81830116. The difference of the latitude, as the 4. degrees, and 6. minutes, multiplied by 15. doth produce or bring forth 61. miles and 30. scruples: which as they may bee wrought and multiplied againe in themselues, they may bee resolued into minutes, and you shall haue 3660. minutes. These further wrought in themselues doe bring forth and shew the second quadrate, which containeth 13395600. The two quadrate numbers also conioyned, doe make 95225716. minutes. The roote of this; that is, 9758. deuided by 60. declareth the space betwéene Buda and Aquisgranum, to be 162. Germain miles and a halfe.

Another.

The longitude of Roome is of 36. degrées, and 20. minutes, the latitude hath 41. degrees, & 50. minutes. The longitude of Ierusalem hath 66. degrées, and no minutes, the latitude is of 31. degrées, and 40 minutes. The difference of the longitudes is of 29 degrées, and 40. minutes. The difference of the latitudes is of 10: degrées, and. 10. minutes. The halfe of the difference of the latitudes, is of 5. degrées, and 5. minutes. The middle latitude, is of 36. degrées, and 45. minutes. The miles answering to one degrée in the Parallel of the latitude nexte following, are 11. and 59. minutes. These subtracted from the miles and minutes of the former eleuation, there doe 9. minutes remaine. These thus founde and knowne séeke the proportional part to bee subtracted, in saying, if one degrée or 60. minutes in this Parallel doe yéeld 9. minutes of a Germaine mile, howe many minutes of a mile doe 45. minutes yéelde or make, which depende to the degrées.

of the Circles.   195

grées of the middle latitude. To know this, multiply 45. by 9 and the product deuide by 60. then will 9. minutes remaine in the quotient. The part proportionall must also bee subtracted, which deducted from the miles and minutes assigned to the latitude 36, as from the 12. miles, and 8. minutes, doe 12. miles, and 2. minutes remaine. By which appeareth, that so many miles and minutes, do answere to one degrée in the Parallell of the middle latitude. This now is as a preparation and entrance, vnto the second working.

To haue therefore the distance of the fore saide citties, multiply first the 12. miles, and minutes, with the difference of the longitudes 29. degrées, and 40. minutes, and they shall bring foorth 356. Germaine miles, and 59. minutes, which 356. miles, that may bee wrought togither with the minutes 59, are to be resolued into minutes, the same is performed, if they bee multiplied by 60. To the same product being 21369. adde the 59. minutes, and they make 21419. These minutes againe multiplied in théselues, do offer the first quadrate, that is, 458773561. Thus you haue the vnderstanding and knowledge of the working of the first place.

After this multiply the 10. degrées of the difference of the latitude by 15. and you shall readily haue the miles 150. to which ad for the 10. minutes dependíng, 2 miles, and a halfe of a Germaine mile, and you shall haue in this second part of the working 152. miles, and a halfe or 30. scruples of a Germaine mile. Which miles, as they may with the minutes bee multiplied togither in themselues, so are they to bee resolued by that 60. multiplied into minutes, which then bring foorth 9120. to which adde the halfe or 30. miles, and you shall then haue the whole to be 9150. minutes: which againe multiplied in themselues doe make the later quadrate to be 83722500. Nowe vnto the last, conioyne these two quadrates, and the whole

D ij.   summe

ſumme ſhall bee 542496061. minutes. The rote of this nūber; that is, 23299. ſeeing it repreſenteth the minutes of miles, deuided by 50. doth then ſhew the ſpace which is betwéene Ieruſalem and Roome, in Germaine miles, to be 388. with a third part almoſt of a mile.

Another.

The longitude of Hamburge is of 37. degrées only, the latitude hath 45. degrées, and 24. minutes. The longitude of Magdeburge hath 29. degrées, and 38. minutes the latitude is of 52. degrées, and 20. minutes. The difference of the longitudes is of 2. degrées, and 38. minutes. The difference of the latitudes is of 2. degrées, and 4. minutes. The halfe of the difference of the latitudes, is one degrée, and 2. minutes. The middle latitude is of 53. degrées, and 22. minutes. The miles aſſigned to the eleuation 53. are 9. and 2. minutes. The miles aſſigned to the degrées of the eleuation following, beeing 54. are 8. and 49. minutes. The difference now of theſe two manner of miles and minutes, hath 13. minutes. The proportionall parte ſubtracted is of 4. minutes; which minutes, let foure be deducted out of the 9. miles, and 2. minutes aſſigned to the eleuation 53. there will then remaine 8. miles, and 58. minutes. Therefore ſo many miles and minutes, doe anſwere to one degrée in the Parallel of the middle latitude. Theſe miles and minutes now found, multiplied with the difference of the longitudes, doe bring foorth 23. miles, and 36. ſcruples. And theſe 23. miles, wrought togither with the minutes; that is, multiplied in it ſelfe, and that reſolued into minutes, to the producte alſo adde the minutes 36. and the whole then ſhall appeare 1416. minutes. This number againe wrought into it ſelfe, doth offer the firſt quadrate, which is 2005056. minutes. After multiply the difference of the latitudes, by 15. miles, and

and the increase shall be 31. miles. These miles againe resolued doe yæld or giue 1860. minutes, which multiplied againe in themselues, doe offer the later quadrante, which containeth 3459600. minutes. The whole summe, that is, the numbers increased of these two quadrats, are 5464656. The rote of the minutes, which is of 2337. minutes, deuided by 60. doth declare the distance which is betwæne Hamburge and Magdeburge, to bee 39. Germaine miles almost.

An easier working. and lesse curious.

This great labour perhaps after the kind, may feare some from the practise of these, and the rather in that this curious or diligent multiplication of the minutes, nædeth not in all or at all times, especially if the space of the two cities doeth not contain many miles, or that the cities offered be but alitle space distant one from the other. For where the distance is great, as of Viteberge & Frankforde, Noriberge and Roome &c. The minutes then neglected, do cause great errour. But if the space be small betwæne the cities giuen, without the acompt also of the minutes (for that seldome in the onely minutes, as are the neare places togither, doe they onely differ) the distaunce then by the onely degrées & miles whole, cannot be found. But if any be minded not so curiously to search the distances of places, then let him or them omit the minutes depending aswell to the degrées of the longitudes and latitudes, as the miles, and according to the instruction of the third rule, the minutes beeing neglected or omitted, you shall then finde without any difficulty the distance of places giuen.

An Example.

The longitude of Franckeforde is of 25. degrees, the latitude is of 53. degrees. The longitude of Viteberge, is of 30. degrees, the latitude hath 51. degrees. The difference of the longitudes, is of 5. degrees. The difference of the latitudes, is 1. degree. The halfe of the difference of the latitudes in whole degrees is nothing, wherefore the middle latitude, is the like nothing. The miles assigned to the lesser latitude, as to the 51. degrees, are 9, multiply nowe these 9. miles with the difference of the longitudes, with 5. degrees, and the increase shall be 46. which multiplied in it selfe, doe offer the first quadrate; that is, 2025. After multiply the difference of the latitudes, that is, one degrée with 15. miles, which 15 miles multiplied againe in it selfe, do produce or bring forth 225. which is the later quadrate. These two quadrates conioyne, and of the increase seek the root, which then declareth the distance betwéene Franckforde and Viteberge, to bee of Germaine miles about 74.

Another.

The longitude of Brunsweeke is of 28. degrees, the latitude of 52. degrees. The longitude of Viteberge is 30. degrees, the latitude of 51. degrees. The difference of the longitudes, is of 2. degrees. The difference of the latitudes, is 1. degrée. The miles assigned to the lesser latitude, are 9. The difference of the longitude multiplied by 9. miles, doeth produce 18. miles, which multiplied againe in it selfe doe produce 324. that is the first quadrate. The difference of the latitude, being one degrée doth make & containe 15. miles, which also wrought againe in themselues doe offer the later quadrate, which containeth 225.

Nowe

of the Circles. 199

Now of these two quadrates conioyned, the rote is of 23. which number is almost the distance of Viteberge, in germaine miles, from Brunsweeke.

Another.

The longitude of Danske hath 39. degrées, the latitude of 62. degrées. The longitude of Noriberge, hath 28. degrées, the latitude is of 49. degrées. The difference of the longitudes is of 11. degrées. The difference of the latitudes, is of 5. degrées. The middle latitude, is of 51. degrées. The miles answering to one degrée in latitude, are 9. The difference of the longitudes, that is multiplied with the 9. doth yéld 99. miles, which againe multiplied in themselues, do produce the first quadrate which containeth 9801.

The difference of the latitudes (being 5. degrées) multiplied by the 15. doeth then produce 75. miles, which wrought againe in themselues do offer the later quadrate which containeth 5625. The increase now of the two quadrates, comprehendeth 15426. The rot containeth 124. And so many are the miles almost, betwéene Danske and Noriberge.

Another.

The longitude of Ierusalem hath 66. degrées, the latitude is of 31. degrées. The longitude of Nazareth hath 67. degrées, the latitude is of 32. degrées. The difference of the longitudes is 1. degrée. The difference of the latitudes, is the like one degree. The miles assigned to 1. degree in the Parallel of the lesser latitude, are 12. The first quadrate doth containe 144. The miles answering to one degree of the difference of the latitude, are 15. The

D iiij.   later

later quadrate, comprehendeth 225. The increase of the quadrates, containeth 369. The rote containeth 16. miles. Now the distance in a maner is so much, betweene Ierusalem and Nazareth. And thus by other examples, may young practisioners excercise, without labour, tediousnes, and paine, to finde the spaces of places giuen, by the degrees of the longitudes, and latitudes.

A demonstration of the third rule.

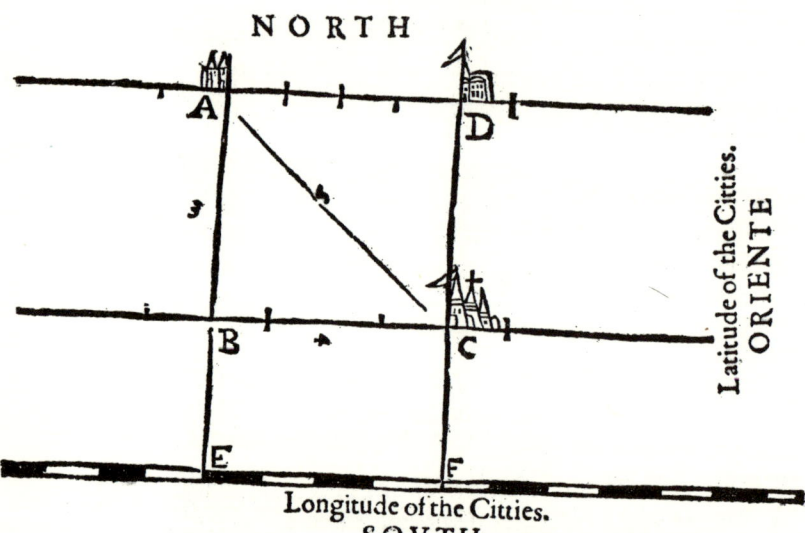

The demonstration of this working or instruction, is taken out of the last proposition of the first book of Euclide, where hee doeth teach and demonstrate, that in the tryangle right cornered, the quadrate which by the line or side drawne and stretching to, maketh a right angle, that is equall in the two squares, which are caused by the sides
con-

of the Circles. 201

containing the right angle. Which that you may easier conceaue and vnderstand, in the page going before is placed an apte figure to this matter, by which, a reason not onely of the third, but also of the rules of the first & second may be practised and declared.

Also there is repeated those thinges, which afore were declared of the Theoricke of the longitudes and latitudes, that the yonger practisers may the readier and easier conceaue the rules hereafter taught. The line E. F. doeth represent the Equinoctiall on earth, lying vnder the celestial Equinoctial cyrcle. The line B. C. doth represent the Parallell; that is, the cyrcle equidistaunt to the Equinoctiall cyrcle, drawne ouer the head or Zenith of the city C. The line A. D. doth represent the Parallel, yea equidistant to that Equinoctiall, drawne by the Zenith of the cities, A. and D. The line A. B. E. doeth represent the meridian, of the proper city or place A. The line D. C F. doth represent the meridian, of the cities C. and D.

The declaration of the first rule.

The two Cities C. and D. agrée in longitude, in that they are vnder one meridian; that is, they bee distant by like spaces from the West. But they haue not alike latitude, for that the City C. is nearer to the Equinoctiall than the City D. by thrée degrées. To haue therefore the distance, or that space betwéene, you shall easily finde the same by the degrées of the meridian.

The declaration of the second rule.

The two Cities A. and D. agrée in the latitude, or they haue one like eleuation of the Pole, in that they are
vnder

vnder one Parallel, and the Zenith of both is by fiue degrées distant from the Equinoctiall. But the longitude of them is not alike; that is, they be not equally distant from the West: for the city A. is more Westerly then the citty D. by foure degrées. So that the distance is to bee gathered and learned by those degrées betwéene, in that Parallell.

The declaration of the
third rule.

The two Cities A. and C. be distant by vnlike spaces, aswell from the West as from the Equinoctiall. For they be vnder diuers meridians and Parallels. The city A. is nearer to the West than the city C. by foure degrées, and it is further distant from the Equinoctiall than C. by thrée degrées. Wherefore by those degrées in which it is nearer to the West and furthest distaunt from the Equinoctiall, must the distance of the two cities A. and C, be sought. For that the space betwéene the meridiane A. B. passing by the Zenith of the City A. and meridiane C.D. stretching by the Zenith of the city C. containeth foure degrées: yet those degrées are not in the great cyrcle, in that those two Parallels doe not deuide the earth into two iust halues, but into vnequall halues: so that of necessity it must follow, that the degrées of diuers Parallels haue vnequall spaces. Wherfore in the third rule are not the miles answering to the degrées of the lesser eleuation taken, except the difference of the latitudes bee small: nor the miles taken, answering to the degrées of the greater eleuation: but the miles are taken answering to the degrées of the middle latitude: for that it lacketh in one part, may be restored in the other. Of the same may the distance in miles be sought, according to the longitude. After this, in that the space betwéene the Parallel A. C. passing by the Zenith of the city A. and the parallel B. C. reaching by the zenith

of the Circles. 203

nith of the city C. containeth thrée degrées, and these are the degrées of the meridian; that is, of the great Cyrcle, where to one degrée doe alwaies and euery where fiftéene Germaine miles answere. So that the distance of those Citties are easily found, according to their latitude.

And in the same by that multiplication of the miles and degrées, the adding of the product, by the increase and extraction of the roote, that the distance of the Cities may necessarily and surely be gathered, is thus demonstrated. That in euery tryangle right cornered, the square which is made by the side, is drawne against a right angle, and is equall to the two squares which are made by the sides containing a right angle. As the quadrate which is made by the drawing of the line A.C. into it selfe, that is equall to the squares, which are caused by the drawing of the line A.B. into it selfe, and B.C. into it selfe: which by Arithmeticall practise may more readier and better bee vnderstood of yong students and practisioners in this maner. First the side A.B. containeth thrée spaces, which multiplied, doe bring forth 9. The line B.C. comprehendeth 4. distances, which multiplied, doe produce or bring for 16. which two squares conioyned, doe make 25 : & the square which procéedeth of the 5. multiplied (which the line A.C. containeth) doe they equate. Guen so in the instruction of finding the distances of places according to the third rule, the difference of the longitudes is represented by the line B.C. but the difference of the latitudes by the line A. B. Therefore as by the quantities knowne of the lines A.B. and BC. is the quantitie of the line AC. attained. Also by the differences of the longitudes and the latitudes of places knowne, and those afore taught being multiplied and increased, the distance of them is easily knowne, which by the line AC. is represented. And in the Triangle and quadrate, is the side (but in the number) named the roote. These hitherto, for the knowledge of finding the distances of places shall suffice.
The

The definition, appellations, diuision,
and offices or vtilities of the Horizont.

THe Horizont called the ender and Cyzcle of the halfe Sphere, is the edge betwæne the light part, that standeth for the same wee sée, and the darke halfe that wee cannot sée of the skie.

The Horizont (as Proclus writeth) is a greater cyzcle, immoueable oz fired, not one and the same euery where, but to each place proper from the verticiall point, and round about equally distant, and deuiding the whole sphere of the world into two equall halfe spheres; of which, the one halfe appeareth in sight to vs, and the other halfe hid vnder the earth.

The description of the Horizont doth Macrobius teach; where he writeth, that the Horizont is after two condicions: the one, extendeth on euery side vnto the firmament and serueth peculiarly as it were for the deuision of heauen, in deuiding iustly the skie into two halues: of which the one appeareth in sight to vs aboue the proper Horizont and the other hid vnder that Horizont from vs. Which Horizont hath his name of the skie, and of the same called the celestiall Horizont: whose diameter (after Macrobius) is as large as the diameter of the eight sphere, which (as he affirmeth) is the furthest and highest parte of the skie, that men can readily sée and discerne with the eie. But the earthly Horizont, in that the same serueth for the sightes onely of the earth and water, and not stretching vnto the firmament; nor that his halfe diameter (as Macrobius writeth) doeth excéede 180. furlongs, which containeth 22. miles,

of the Circles. 205

miles, and ¼. So that the whole diameter after his account, is but 45. miles in length. Which if any man stand vpon an euen or plaine ground (or els on the sea) may see round about him 22. miles & a halfe euery waies. Which rounde compasse of the whole Horizont (after Macrobius) doth containe 141. miles, and ₇. parts.

A comparison, that as the meridian is an immoueable cyrcle, euen so is the Horizone: for if the same were moueable, it woulde not crosse the meridian at right angles: and vnto these should be imagined, that if it were moueable, in each day the same would mooue with the meridian cyrcle.

The appellations aud diuers names of the Horizont.

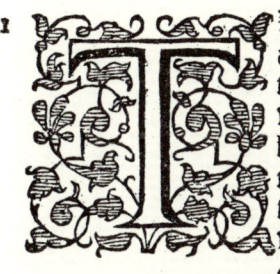

1 His circle is called the Horizon, as it were the cyrcle deuiding the halfe spheres, or of the græk worde *Orizomai*, which in English signifieth to define, determine, and set out, in that the same defineth the parte of the worlde sene. Or of *Oros* or *Orion*, that is the bound or ender.

2 It is named also the gyrdle, or the cyrcle of rising.

3 Macrobius calleth the Horizone that bounde of heauen that is sene aboue the earth (lib. 1. cap. 15.) In that it is the end or bound seperating the neather halfe Sphere from the vpper. And of him also called the edge of the halfe sphere. And Alfragnus called it the cyrcle of the halfe sphere.

4 The Horizone also is so defined of his office, in that his office is to deuide that part of the worlde in sight, from that hidde vnder the earth. Whereof it is not vnworthily cal-

called the ender, seeing it permitteth nor suffereth any to see but the halfe sphere at one time, and therefore is called of some, the cyrcle of the half sphere, as afore taught. This cyrcle is alwaies vnderstood to be described by the verticiall point, in that as the verticall point is changed, euen so likewise is the Horizone.

The Horizone is deuided after twoe sortes; first into a right and thwart: secondly, into a sensible and rationall Horizone.

The Horizone of the right sphere is called right or right cornered, aboue which neither of the Poles of the worlde is eleuated, which they haue whose Zenith is vnder the Equinoctial, or dwell vnder the Equinoctiall. Their Horizone is the cyrcle drawne by the Poles of the worlde, which deuideth aswell the meridian as the Equatoure at right angle, through which rightnesse it obtaineth that name, that it is called the right Horizone.

The thwart Horizone as of the thwart Sphere, from whose plaine the Poles of the worlde be distant, the one is then raised aboue the Horizone, and the other depressed and hid vnder the Horizone. Or thus, the Horizone is called thwart or declined, when either of the Poles of the world is eleuated, which they haue which dwell without the Equinoctiall, whether they dwell Northerly or Southerly. And their Horizone crosseth the Equinoctiall at vneuen and thwart angles. And attaineth also the name of a thwart Horizone, through the thwart angles, which it formeth or maketh with the Equatoure. There is also one right Horizone, as there is one simple & right sphere, but the thwart Horizone is many waies changed toward the Poles of the worlde, through the standing and place chaunged on the earth. For the standing is so much the thwarter, as the sphere of the worlde is caused declining, & by how much either of the poles of the world, is drawne and raised higher. So that to it (by the obseruers of the stars)

of the Circles.   206

ſtars) is the name giuen whereby it is called thwart. And note this that the zenith or point ouer the head, is alwaies the Pole or the Horizone: which Pole is here not taken for the celeſtiall point, vpon which the celeſtiall mouer or any other cyrcle is drawne, in that the Horizone is immoueable, as was afore taught: bu taken for the pointe raiſed, which is the Center of any cyrcle, as here by this figure

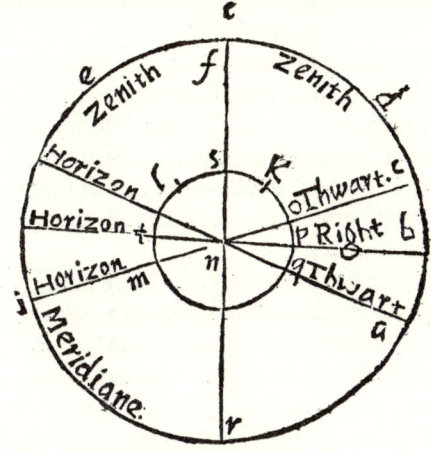

doe all the notes appeare vnto the eie, in which F. H. R. repreſent the meridian cyrcle, F. R. the halfe of the equatour H. N. B. the right Horizone, croſſing aſwell the Equatour in the point N. as the meridian in the points H. B. at right angles, F. is the verticiall point of the dwelling vnder the Equatour, in the point S. The letters K. O. Q. doe repreſent the earthly Globe; and he which dwelleth without the Equatour in the point K. hath his top the pointe D. in the meridian, and G. N. A. the thwart Horizone, but dwelling in the earthly Globe, in the point L. hath the verticiall point E. and the Horizone thwart I. N. C. But if any vnderſtand or mean that L. K. Q. is the Orbe of the earth

whoſe

The second Part

whose Center is N. and shall soone see by the Horizonts or ends that that Orbe is deuided into two equall partes, as into that seene, and that not seene: and by the right Horizont H. B. the Equatour R. F. in the point N. at right angles, whereof it is called the right horizonte to be deuided, and the others at thwart angles, whereof they are named thwart. Of this deuision hath been sufficiently intreated, in his proper place. This also is to be noted, that the horizone is two waies changed and varied. First, that the Cities and other places are situated or standing either toward the East, or toward the West, vnder one Parallell. The second, that the horizone is either varied toward the South or towarde the North, and are situated vnder one meridian, and diuers Parallels. These of themselues for the right vnderstanding the longitudes and latitudes of places, are manifest.

Further that the eleuation of the Pole (that is the arke which is betweene the Pole of the world and the horizon, is equall to the distance of the verticiall pointe or Zenith from the Equatour, shall appeare and bee made manifest

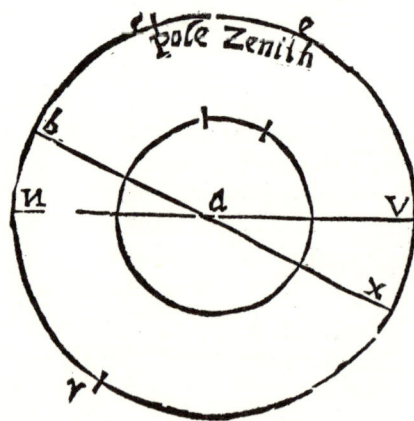

of the Circles.

in this maner. As first that E B VX. is the meridian cyrcle BAX. the Equatoure, NV. the thwart horizone, the Pole of the worlde raised aboue the horizone, E. the Zenith, A. the Center of the worlde, R. the Pole antarticke so much depressed, the arke RNC. the eleuation of the Pole from the horizone NV. giuen, EX. the latitude of the region or place being the distance of the top pointe E. from the Equatour BX. And that this is equall to the arke NO. as to the eleuation of the Pole C. from the horizon NV. giuen, which shall be demonstrated in this manner. That C. the Pole of the world, is distant from the equatour; BX. by a quarter of the meridian, and the like the verticall pointe E, is distant from NV. the horizon, by a quarter. But the quarters of the equall cyrcles are alike equall, for XC. and NE be the quarters of the equal cyrcles: therefore in that they be alike equall, is CE. the common arke. If now by the common conceiuing and imagination of mind, that from the equals they bee equall, &c. then from either quarters XC. and NE. the common Arcke C E. is to be deducted, & the remainer shall bee equall; that is, XE. which to that NC. ought to bee. And seeing the latitude of a place is no other, as by the former words appeareth, the the distance of his Zenith from the equatoure, that readily hauing the eleuation of the Pole, the latitude or distance of the place from the Equatour shall soone be attained. By which the eleuation of the Equatoure aboue the Horizone in the meridian cyrcle, as the arke VX. is, sheweth no other then the complement of the latitude or eleuation of the pole is readily atained, if you deduct that complement out of 90. deg.

The sensible Horizone is a space of the earth defined by a compasse rounde about, which the sight of the eie attayneth and comprehendeth in a plaine and euen field. Or thus, the sensible Horizone is that which the eie perfectly seeth, and describeth according to the bounde of sight, and called of some the artificiall Horizon, and that for the same

P j.   tificial cause, that which is contained by sight, is by a certaine similitude agréeing with the artificiall day. And as the artificiall day is so named, for that artificers doe especially worke in it, euen so the like is the horizone named artificiall, in that towers, foretresses, and castles in time past, were built like the horizone.

The diameter of this horizone (after Macrobius) which nearer agréeth to a truth (then either Proclus or Albertus) as afore was taught, is of 36. furlongs, to which almost foure Germaine miles answere, and 22. English miles: and so far on a plain and euen ground not hindred by hils or thicke mists, may a man fully sée. And in the same space the imbossed rounds of the earth, being without hils, is increased, and groweth to 250. féete, or 125. cubits: so that this horizone is not sodainly changed, nor in a short space. Therefore of necessity must ensue, that those which are distaunt by a lesser space then 360. furlongs, to sée alwaies some part of the earth common to both. But those which are distant by many spaces, doe comprehend diuers compasses by sight of the eie and diuers horizones.

The rationall horizone is that which afore was described, that the same is a greater cyrcle, lying by the edge of the earth, and reaching round about vnto the skie, and deuiding the celestiall Orbs into two equall halfe Spheres, as the one halfe in sight, and the other hid to vs. Although the plain vpper face of the horizone passeth not by the center of the earth, yet by the edge of the same, through which we sée and obserue the celestiall bodies, that rise aboue and set vnder it: so that they euidently shew, that the same deuideth heauen into two equall halfe spheres, as aboue remembred. For in euery moment, doe sixe signes of the Zodiacke appeare aboue the earth, as in the night to the eie may be numbred and noted, that sixe signes set vnder the earth, and be gone out of sight. This is also called rationall, séeing the eie cannot descerne vnto the highest heauen, nor aptly frame this diuision of heauen into two e-

of the Circles.

quall halues: yet the mind by examining, gathereth and concludeth, as by a perseuerance passing before, and in the shewing of the starres that rise and set, and in considering the tarriances of them in either halfe sphere. This besides is called the artificiall horizone, in that by the benefite of the astronomicall art, it was inuented.

Or thus not much agréeing to the former, the rationall horizone (which of some is named natural) and according to the mind of Ptholomie, Cleomedes, and Proclus, belongeth vnto the Sphere of the fired stars, and reacheth euen vnto the same Sphere, and deuideth heauen into equall halfe Spheres, the one halfe appearing aboue the same circle, and the other halfe not appearing, hid vnder it. Such a maner of imagining is not in vaine, nor without cause determined and deuised, séeing that men in the night and in a cleare season, standing on an euen grounde, may sée stars arise vnto sight in the East, which a litle before appeared not to the sight: and those after drawne by the first moouer vnto the West horizone, that began to go downe be set, and doe not after appeare. By which they concluded that there is a cyrcle in heauen, deuiding and ending matters in sight from those not séen. So that they nothing doubted to call this cyrcle the rationall horizon (which togither with the vpper face by the center of the earth stretched round about vnto heauen) and by the foure quarters of the world, as East, West, North, and South, deuided things séene, from those not séene. And a great helpe it giueth vnto this imagination that the earth is perfect round and imbossed, in that of a Globe through his imbossing can be séene but the halfe at a time.

This also yéeldeth a helpe to reason, by the appearances in the celestiall bodies, although our sight cannot attaine vnto the starrie sky, nor fully descerne heauen, although a man earnestly looke vp and behold it: yet doe we sée stars, whose light extend vnto our eie. As by this example may euidently appeare, of that royall ſtar named the heart of the Lion, which in our time is in the 22. degrée almoſt of Leo. And the ſtar ſtanding on the left buttocke of Aquarius in the 22. degrée almoſt of the ſame ſigne, that is diametraly or right againſt one the other ſituated. Which doe on this wiſe, that as the one appeareth aboue the horizone, the other is hidden vnder it, et e contra. So that as the one riſeth, the other ſetteth, and on this manner doe they continually. Of which reaſon it is concluded, that a certaine cycle deuideth heauen into twoe equall halues, and do part (as afore taught) the things ſéene, from thoſe not ſéene. Although the tariance be but ſmall, in that this ſtar appeareth a very ſmal while aboue the earth, through the ſame, that this ſtar of Aquarius is Southerly from the eccliptike line, it greatly forceth not. The like examples may be applied of the ſuperiour planets, when they be ſituated or appeare oppoſite in heauen, as they alſo may be euidently ſéen, in the oppoſition of the ſun and moone, when they bée ſéene neare to the Eaſt and Weſt horizone, and where the moone is neare the ſuns way.

The diameter of the rationall horizone, although the ſame cannot be found nor comprehended, through his excéeding diſtaunce by exteriour ſence and iudgement: yet reaſon it ſelfe iudgeth, that the ſame may extend vnto the ſtarry ſky, whoſe ſight from that not ſéene it doth deſcribe and the ſame is of 32655932. Germaine miles, and 20. minutes, which diſtance by the outward ſenſes, is iudged as infinite.

The Pole of the rational horizon, is the verticall point. For it is diſtant by a quarter of the greateſt cycle, that is, 90. degrées, from the compaſſe round about of the horizon, yet not to all places ſerueth one horizon, for that as a man changeth place and country, euen ſo ariſeth a newe horizon, whether ſo euer he trauaileth. And new horizons alſo appeare and happen, if a man either trauaile toward either

of the Circles.

ther of the poles of the worlde, or in right line toward the East and West, and the like vnto diuers quarters, as into the North, the East, or West, or contrariwise iourneying by the opposite course, the Horizones vary and change.

And if the places bee either situated partly toward the East or West, and partly toward the South or North, the horizones there decline and varie them partly toward the East or West, and partly toward the south or North: which hapneth, by reason that the City is not vnder one Parallell.

And Cities or countries situated vnder one meridiane doe vary their horizons directly; either toward the South or North.

There be as many horizons, as there be meridias. And for so much as that of all places cannot bee one manner of Zenith, therefore cannot one Meridiane serue for all places. And seeing the Pole of the Horizone is the Zenith of it, which is in the Meridiane, and that to each place belongeth a proper Zenith, and a proper Meridian, it followeth that to each place belongeth a proper Horizone.

Toward the Poles by the chaunging of places are the horizons chaunged, and the diuers eleuations of the Pole by a certaine occasion caused: also they euidently declare a like alteration to bee caused in the respect of the opposite quarters of the East and West, and doe procure and cause diuers beginnings of the daies and nights, insomuch that the starres generally appearing and seene, doe by order of times and in sundry places, arise and set in the West, and hide them vnder the Horizon. For the same maner of Eclipse, which is seene at Arbela (after Plinie) in the fifte houre of the night, to them of Carthage it appeareth in the second houre: so that the sun sooner setteth to them of Arbela by three houres, then to them of Carthage. Therefore

P iij.   the

the horizon of Arbela is much further distant into the East then the horizon of Carthage.

The same rationall horizon (as it were on the plainesse of the earth) drawne and streached vnto the sky, doeth the meridian extend to it downward, and deuide the same into twoe halfe cyrcles: of which the one declineth vnto the East, and therof called the East quarter, and the other vnto the West, and of that named the West quarter.

And the diuers places of the suns rising and setting, doe sundry wise deuide either halfe cyrcle. For the Equinoctiall rising, and the Equinoctiall setting, (which are points of the horizon, that the sun in the equatoure placed, by rising and setting passeth) doe parte and deuide either halfe cyrcle into equall quarters. And with these points do the Poles of the meridian ioyne.

And either quarters do the other two (as the rising and setting) deuide into two vnequall arks. For of the twoe Northerly quarters, the same which tendeth and looketh vnto the East, is the solsticial rising, and the other the solsticiall setting. But of the twoe Southerly, the Easterly doth the winter rising deuide, and the Westerly doeth the winter setting part. But by what space these risings and settings may differ and be distant from the former middle in euery horizon, and in the largenesse of rising doth Ptholomie instruct in that eleuation of 40. degrees, and fifteene scruples.

Of the shadowes which the sun arising and setting in these points of the horizon causeth, is worthie to bee considered and noted, in that the Equinoctial shadowes (which through the sunnes rising and setting in the Equinoctiall pointes are caused) doe fall and extend in straight maner. But the other shadowes not in the same condicion or not in straight line doe fall, but that the solsticiall shadowes in the rising, with the winter shadowes in the setting, and contrariwise the winter shadowes in the rising, with the

Solsticiall in the setting doe fourme and make right shadowes.

The offices or vtilities of
the Horizon.

1. His circle (like as al the others) so that nothing in heauen is friuolous and of a vaine imagination hath many vtilities. First it deuideth the whole heauen into two equall halfe spheres.

2. It declareth which starres be of continual appearance, and which continually hid vnder the horizon: which doe set, and which doe arise aboue the horizon. So that it appeareth, that the stars consist in a triple defference, as that certaine do arise and set, certaine neuer appeare aboue the horizone, and certaine continue and be alwaies aboue the horizon.

3. The horizon therefore is caused of the habitude, as well of the right, as the thwart sphere.

4. The rising and setting of the stars are applied vnto the horizon, by which settings and risings, the discriptions of times are chaunged, and it also declareth the degræ of the Zodiacke, with the which each starre riseth and setteth.

5. The horizone sheweth the rising and setting of the signes of the Zodiacke, the exaltations, or eleuations of the pole and the equatoure, the latitudes of places, to the largenesse of rising, which is the arke of the horizon to the stars or points of the eccliptcke and equatour, arising togither, included with the beginnings of the twelue houses of heauen.

6. By the office of the horizone, at any time wee may learne

learne and knowe the quantity of the artificiall day and night: and likewise procureth or sheweth the iust cause of the inequalitie of the artificiall daies, it doeth also declare the rising and setting of the sun. For as the horizons, according to the eleuation or depression of the pole, are varied: euen so are the verticiall daies in themselues caused vnequall, yea in those points of the Zodiacke.

7 By the benifit of the horizon (the sun shining) we attaine and come each day vnto the knowledge of the vnequall houre of the day.

8 It sheweth to vs the elongation of the stars from the rising and setting, which the astronomers call the largenesse of the rising and setting, or the Zenith of the rising and setting.

9 By this cyrcle we learne how much the rising aswell of the stars, as the other points of heauen, is distant from the true and the Equinoctiall rising: that is, in the same are the latitudes of the stars accompted from the equinoctiall, and also their risings and settings.

10 It manifesteth the degrée of the Zodiack, with the which the purposed star riseth and setteth.

11 It indicateth the stars, or the celestiall images that be continually in sight, or alwaies hid.

12 It maketh manifest the risings and settings of the signes of the Zodiacke. It doth likewise make distinction, betwéene the Sun and Moones Ecclipses, séene as well aboue the horizon, as not in sight.

13 It helpeth and furthereth much vnto the finding of the latitude of a purposed place, whereof (through the benifite of this cyrcle and the meridian) may the distaunces of places be certainly found.

Of

of the Circles.

Of the verticall Circles.

Esides the former cyrcles at large mentioned, are there other cyrcles which shall here bee vttered and taught, as in an apte place agræing to that aforesaide, which bee these: the verticall cyrcles; the cyrcles of the positions, and of the 12. houses. Of these in order shal here bee written (as the necessary matter offereth) seeing a speciall part of astronomie dependeth of them, and the whole composition of the celestiall instruments seemeth likewise one of them.

First the verticals, be cyrcles which from the top of any place giuen, are drawne vnto each part of the horizon, and deuide the vpper halfe Sphere in sight into so many partes, as the Horizone is deuided; and all concurre and meete aboue in each verticall pointe or Pole of the Horizon. To the number of these, is the meridian adioyned. These cyrcles, are likewise vnderstoode and noted immouable, as the meridian and Horizone; that is, they are not drawne about with the first mouer, as the Zodiacke, the Equatoure, the Colures, and the other cyrcles infixed to the first mouer.

But for a more euident declaration of the former wordes, vse this Figure here described: whereas a e g d. represent the Meridiane; e d b g. the twart Horizon: a f. the Poles of the same. And from the verticall point or pole of the horizon, vnto e g d. the halfe of the horizon, which is deuided into equall partes: and the quarters of the Circles drawne to it (which are the partes of the verticall circle) which if they be wholy described, doe concurre and meet in the opposite pole of the Horizon f. To these Circles

The second Part

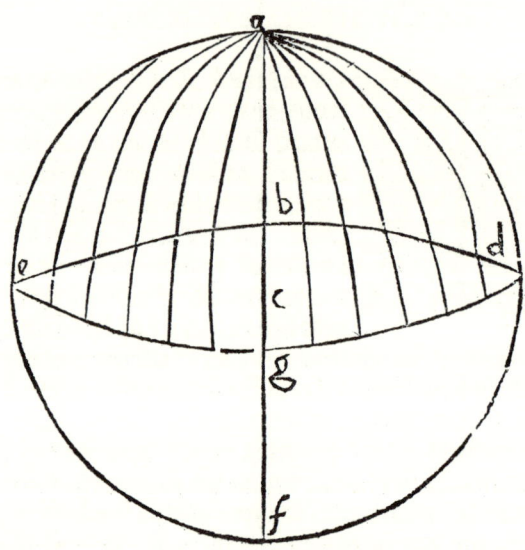

cles (as is afore mentioned) is the Meridian added. Further, the letter c. doeth signifie the North, d. the South, b. the East, and g. the West. By which appeareth, that the horizon from the Meridian Circle a e f d. and a c f. (by which line the Circle is represented, passing by the verticall point and the true or Equinoctiall rising) is deuided into foure quarters, b c, b d. as the two quarters orientall, g e, g d, the occidentall. In the orientall halfe of the Horizone e b d. is e b. called the Northerly quarter, and b d. the Southerly. And in the Westerly halfe e g d. of which e g. is the Northerly quarter, and g d. called the Southerly.

The

The Circles of the Altitude.

THe Cyrcles of the altitudes bee those, which are equidistantly described about the toppe of places. As the verticall cyrcles doe deuide each of these cyrcles into 360. degrées, euen so doe these deuide a quarter of each verticall cyrcle into 90. degrées. So that none of the altitude cyrcles is greater then the horizone, nor lesser than that which is imagined and vnderstoode to be described about the verticiall pointe. The especiall office of these cyrcles is, that aswell the altitudes of the fixed stars as the Planets, may bee measured and knowne, as the fixed stars aboue the horizone: by which altitude or eleuation, the times; that is, the houres are knowne, and the places of the starres, as may appeare in tables made for that onely purpose. Séeing then it cannot be (and that through the roundnesse of heauen) but that any star giuen or supposed vnto the motion of the whole, is imagined by his altitude to be distinguished in some Parallel: therefore is the altitude of the star or of any other celestiall point, the arke of the verticiall cyrcle, drawne by the Center of the star, contained betwéene the horizon and the star giuen, which (as afore written) is distinguished of the said parallell. The méeting and ioyning togither of these cyrcles with the verticals, is not moued, but at the motion of the verticall point; which is none other, then the pole of the horizon, from which all the parallels of the altitudes, are imagined to be described by equall distaunces. But this (in mine opinion) is not to bee ouerpassed; that is, that any star, when it shall be equally distant from the meridian, either hath or may haue the same altitude

from

220　*The second Part*

from the horizon, as to the eie is offered in this figure folowing.

Where a b d c. is the Meridian: b e c. the greatest halfe of the Paralels of the horizon: f g. the leaft: b. the North, and c. the South: a. the point of the top: o. oꝛ q. the place of the ſtarre giuen, by which a o k. oꝛ a q n, the verticall Circle paſſeth, and the like doth the Parallell p l k. The Arke k o. oꝛ n q. is affirmed to be the altitude oꝛ eleuation

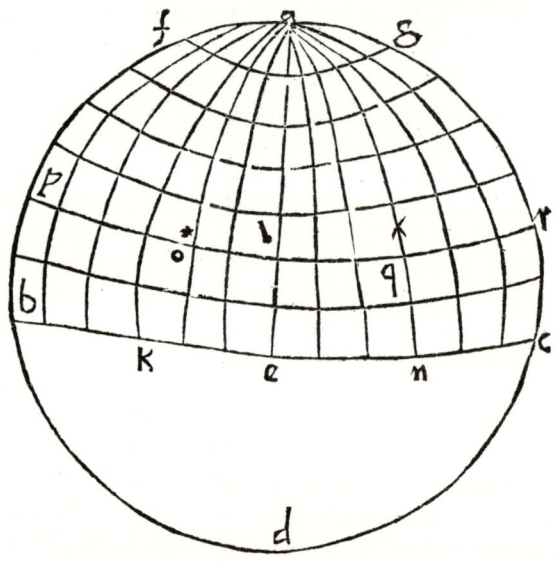

of the ſtarre from the horizon, that endeth at the parallell p o l r. and p o. is the diſtance of the ſtarre from the nonesteed a p d c. Now when the ſtarre (by the motion of the principall) is drawne vnto the point q. in which when the ſame ſhal be, it will be equally diſtant from the Meridian Circle: wherefoꝛe thꝛough the equal diſtance of the parallels, of which they be named, ſhall the Arke o k. bee
equall

equall to the Arke n q. Of this procœdeth and is caused that in the howers equidistant from the nonestead, as is the seauenth houre before noone, and the fifte houre after noone: likewise the eight and the fourth, the ninth and the thirde, the tenth and second, and so of the rest. The sunne obtaineth equal eleuations aboue the horizon. This much auaileth in the composition or making of dials, and giueth great light and breuity to the same practise, as may appeare elsewhere: but the verticall Circles in the solyde Spheres and Globes, by one quarter of the Circle, depending of the verticall point vnto the horizon (diuided into 90. degrees) is declared.

The houre Circles.

IN that the whole worke of dialles dependeth vpon the knowledge of the houre cyrcles; it is therefore requisite and necessarie to entreate fully of the cyrcles distinguishers of the houres, or at the least vtter a brief instruction of this. First you shall vnderstande, that the Equatoure onely, which (as afore taught) the sunne beeing either in the beginning of Aries or Libra, is regularly moued, aswel in the right, as thwart Horizone; and thereof is alwaies the one halfe aboue the horizon, & the other halfe hid vnder the Horizon. Through this his equall motion or regulare motions, is it iudged worthy and laudable: seeing by it the equall houres (as well by day as night) are attained and had. And this conceaue, that there are twelue greater cyrcles vnderstode, which crosse the Equatoure at right angles, and passe by both the poles of the first mouer, from which the said equatour, is distinguished into 24. equall parts, which are called

led the diſtances or ſpaces of the houres, in that each be diſtant from other by 15. degrées. For they deuide the verticall, the Zodiacke, and the horizone into 24. partes, but vnequally: at which Poles the nearer partes to them are narower then thoſe which be and draw nearer to the equatour. And that theſe may clearer and perfecter be vnderſtode, imagine your ſelfe to bée vnder the equatoure; that is, in the right Sphere: in ſuch a ſtanding ſhall the halfe meridian Cyrcle bee the line of the twelfe houre, and the halfe horizontal circle, the line of the ſixt houre before noon: and the other halfe of it, the line of the ſixt houre at after noone. By which imagination firmely conceiued, may a man imagine betwéen the halfe horizontall cyrcle, and the halfe meridian cyrcle, to be other fiue halfe cyrcles firme and immoueable, which are not mooued but as the verticall point is moued, being diſtant each from other by an equall diſtance, as by 15. degrées of the equatour. The firſt after the horizon, is applied to the ſeuenth houre, & ſo forth of the reſt. And in like maner betwéen the meridian halfe cyrcle, and the occidentall horizone are other fiue cyrcles vnderſtode (according to the fourmer deuiſion) and that which followeth the meridian, ſhall be applied to the firſt houre, that which next foloweth to the ſecond houre, and ſo forth of the others. Beſides, imagine the ſun to aſcend from the horizon, and when he ſhall be come vnto the firſt halfe cyrcle from the horizon, then ſhall he ſhed a ſhadowe furtheſt weſtward, and being drawne vp vnto the ſecond, ſhal make a ſhorter ſhadow, and the ſhadow ſhal alwaies (vntill the ſun bee come vnto the halfe noneſtǽde cyrcle, where he ſheddeth or ſendeth a ſhadow) plum down right to the earth: but deſcending from the Nonſtǽde vnto the Weſt, the Sunne cauſeth then the like ſhadowes contrarie.

  Further conceiue, that the Exe-trée of the worlde, in whoſe poles (as is afore taught) all the houre cyrcles méet to

of the Circles. 223

together in one, doeth performe & expresse the same, which the foresaid cyrcles taught: as by the sun dials the like is readily understood and knowne.

Which this Figure further explaineth, where a b d e. represent the Meridian, a i d. the right horizon, b i e. the Equatoure, d. the Northerly pole, and a. the Southerly pole. In these two poles doe all the hower Circles meet, as the same here appeareth unto the eie, and the letter b. is the vertical point. The distinguishers of y howers by the nu-

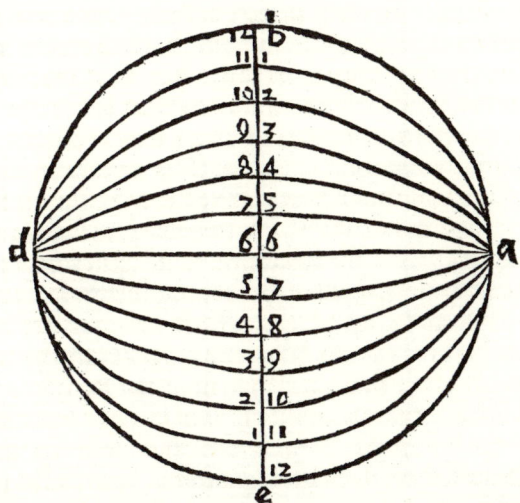

bers added are manifest. For a b d, is the halfe Meridian circle, as already saide: the letters a i d. doe represent as wel the halfe Oriental as the Occidental circle of the horizon: a b d i. is the halfe of the World above the Earth a e d the other halfe hidden. The eie of the beholder, imagine to stand in the point i. and each of the houre Circles above the horizon understood and taken or used twice togither. To conclude this sphere doth offer and teach clearely and
plainly

plainly all the former. So that the pole articke with the rest of the cyrcles must be raised aboue the horizon, and the halfe cyrcles of either sirt houre, seuered or deuided from the horizon. Of this ensueth, that the equatoure leaueth the verticall point, and howe much the Northerly pole is raised aboue the horizon, somuch doth the equatour depart from the verticall pointe (as afore in the proper place is aptly demonstrated) & howe much one quarter of the halfe cyrcle of the sirt houre is raised togither with the Pole, so much the other quarter with the opposite pole is depressed and standeth vnder the horizon. Of this proceedeth, that they crosse one the other in the East parte, and that in one and the same point, the Equinoctiall, the horizontall cyrcle, and the halfe Cyrcle of the sirt houre of the morning. These throughly learned and vnderstoode, and the sphere applied to the materiall with any houre cyrcle, by which the harder or more curious matters are made manifest and plaine; you shall then readily see, that the sun whiles hee runneth in the Northerly signes, doeth sooner come in the morning vnto the horizontal circle, then vnto the halfe cyrcle of the sirt houre in the morning: but the sun running in Southerly signes, he then causeth the contrary; that is, he attaineth or commeth sooner vnto the halfe cyrcle of the sirt houre of the morning, than vnto the horizon. And of this ensueth, that the nights here are longer, but the daies there bee the longer. The arke furthermore contained in the Parallell cyrcle to the Equatoure, and passing by the suns place (betwéen the horizon and the halfe cyrcle of the sirt houre) is the difference betwéene the equinoctiall day, and purposed day, whatsoeuer or how much the same be, which is worthily to be noted. Besides these, it greatly auaileth to vnderstand in these thrée cyrcles, the equinoctiall, the horizon, and verticall cyrcle, that the vpper faces ending at those cyrcles, the equinoctiall truely by the same receiueth and causeth a deuision that the vpper face also is

sup-

of the Circles.

supplied and placed vnder, and receiueth and maketh the like; that is, equall. But the vpper face which is placed and standing vnder the horizone, doeth receiue and make an vnequall deuision, euen the same that the horizone is, which of the houre cyrcles is vnequally deuided. And the like also may be gathered and iudged of the deuision of the vpper face of the verticall cyrcle, which (euen as his cyrcle) is vnequally deuided. But that these(for breuity) may readier and plainer appeare, conceiue this figure following demonstrated, which without long circumstance

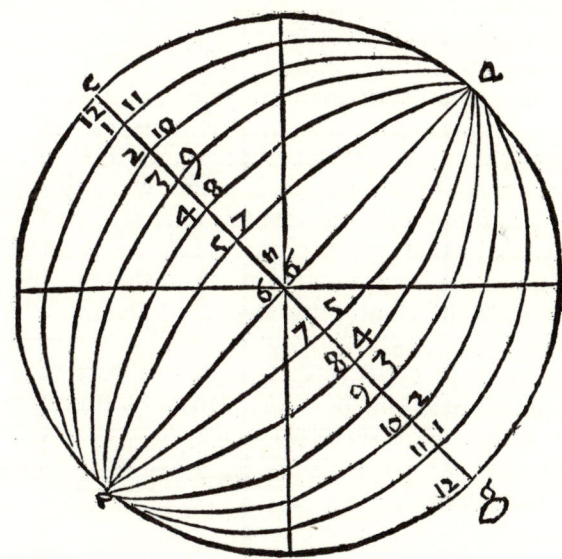

of words, doth euidently set forth that which is set down before. The letters b h f. represent the Meridian Circle. g n c. the Equatour, h n d. the twart Horison, b n f. the verticall circle: a. the northerly Pole, e. the Southerlie pole:

pole, h a. the eleuation of the pole aboue the horizon, d e. the submersion or standing vnder of the pole, right against the point n. is the pointe in which (as afore taught) the thwart horison h n d. and the equatour c n g. and the halfe cyrcle of the first houre of the morning a n e. togither with the verticial cyrcle b f. doe crosse one the other in the East. If you consider the deuisions of the halfe cyrcle houres in the Horizone h n d. you shall see them to bee togither vnequall, as nearer to the East point n. but larger to the point h; that is, narower to the meridian cyrcle. And euen the same hapneth in the verticall cyrcle b n f. that is, that the halfe houre cyrcles toward the Equinoctial Horizon from the Nonesteede by their largenesse encrease, and from the Horizone toward the Nonsteede in the same verticall circle doe decrease: and howe much the more the Northerly Pole is eleuated aboue the Horizone, and the opposite depressed, so much the narower doe the aforesaide deuisions alwaies goe or come together. The now being throughly learned and known, doth bring and yeeld a great commodity in the making as well of the horizontall as murall or wall dials: euen as those dials which are made on walles looking into the South, wheather they hang directly or thwartly, according to their distinctions, from the diuision of the verticall cyrcle by the halfe houre cyrcles: but the Horizontals are, from the diuisions in the horizon by those halfe cyrcles. These hetherto briefely touched for the vnderstanding and knowledge of the houre cyrcles shall at this time suffice.

The Circles deuiding the twelue houses of Heauen.

NOw resteth to entreate of the cyrcles distinguishers of the houses, and the cyrcle of the positions: but first I will write of the distinguishers of the houses. As there are

are sixe cyrcles that are imagined of the astronomers, by which heauen is deuided into twelue parts; among which are the Horizone and meridian, whereby the whole is deuided into foure equall parts: and those twelue parts, are (of the astronomers) called mansions or houses. But as touching the constitution and forming of the celestial houses, there are sundry old and late opinions, but whether opinion is the worthier, or to be the rather allowed, is not here mente to bee stœde vpon, nor aptely belongeth to the matter I entreat of, so well as in the proper place is agréeing: yet certaine, and especially the auncient, which were Campanus, a singular mathematician and astronmer, deuided the houses by the fiue cyrcles of heauen, méeting and ioyning at the Poles of the worlde; from which they deuided the whole heauen (togither with the meridiane) into twelue equal houses. But for a better and readier instruction, they formed and drewe them in this maner. After the foure principall quarters or angles of heauen were drawne, and that the right ascention of the middle of heauen was had, then were the partes of the Zodiacke diligently considered (that occupy as well the Easterly as the Westerly Horizone) and then were the right ascentions sought of those partes: which being done, the constitution and making of the two houses in the Easterly part of heauen, was the right ascention of the mid heauen, deducted from the right ascention of the Horizone: and the remayner, was distributed into thrée equall parts. In the bound of the first part (in accompting from the Nonstéed toward the East) was there imagined a cyrcle for the beginning of the eleuenth house: but in the bound of the second parte (from the Nonestéede) was the beginning of the twelfe house placed. After in the bounde of the third parte, from the Nonstéed, was the beginning of the first house drawn: and the like was wrought and done in searching for the 2. Westerly houses, as the ninth and the eight house. For

Q. ij. they

228   *The second Part*

they deducted and subtracted the right ascention of ẏ west part, from the right ascention of the mid heauen or noone-stead, and the remainer or rest (as afore taught) was distributed into three equall parts. After that in the ende of the first portion (from the nonstead towardes the West) the auncients constituted or placed the bound of the ninth house, with the circle comming from the poles of ẏ world: and in the bound of the second portion, was the beginning of the eight house formed. These attained, the degrees and partes of the degrees of the Zodiack answering to ech arkes of the Equatoure, were sought in the Tables of the right sphere: but the houses standing vnder, were defined and made like to their opposites. And seeing this maner of forming the houses is vnperfect, therefore shal here no further be taught of the same.

But the other Astronomers, as Campanus and Gazu-

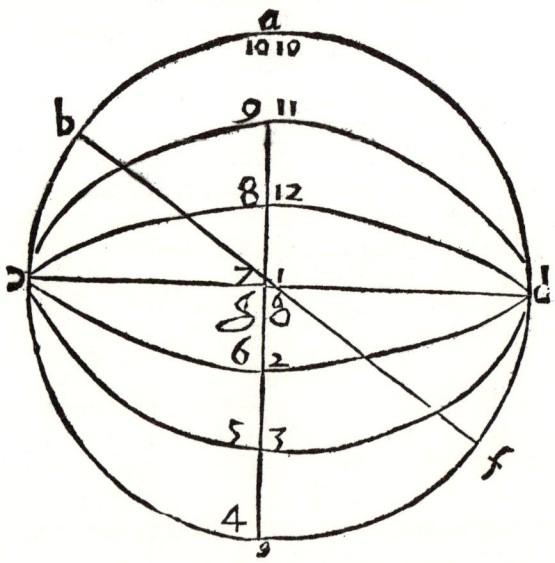

lus

of the Circles.

lus, doe otherwise handle this matter, which with the 4. circles in the sections of the Horison and Meridian méeting and ioining, diuided (togither with the Meridiane and Horison) the whole heauen into twelue equall parts: which equalitie in the circle passing by the Zenith, was the equinoctiall rising considered, as the same may more plainer and euidently appeare in this figure here demonstrated.

In which a e. is the verticall circle, crossing a d e c. at right angles: f g b. the equatour: d g c. the horison, d. and c. be the points in which the distinguishers of the houses concurre and méet; which also do make equall distinctiōs in the verticall circle, and thereby be the houses noted and diuided.

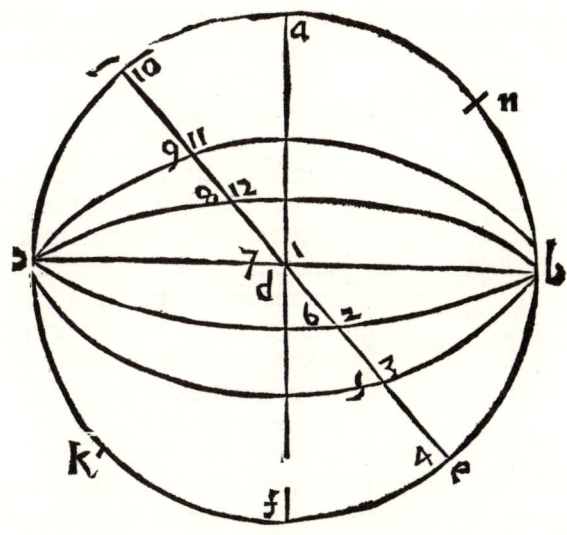

But the later Astronomers, moued by the authority of the incomparable Mathematician Regiomontanus, inuēted

ted and deuised another order of the houses, more agreeing to reason than the former. For they deuided the quarters of the equatour, comprehended betwæne the horison and noonstead, into thrée equall spaces, and by each section they imagined great circles, ioyning in the sections of the Meridian and horison, as the former, Although all these are plainer and more euidently taught and known in the materiall Sphere, yet we thought good to speak somwhat (as our possibility serueth) in plaine forme.

Wherefore graunt that a f c. is the Meridian, a. the Top, n. the Northerly pole: k. the Southerly pole, b. and c. the points of the sections of the horison and Meridian, where the distinguishers of the houses concurre and méet, which also are imagined by the equall distinctions of the equatour e i l. as to the eie sufficiently appeareth, that b i c. is the horison circle, d. the easterly point or rising of the equatour, from which the first house taketh his beginning.

The Circle of position.

ALL these Circles being set down, the Astronomers notwithstanding do write of another Circle, whose vse and office serueth to great purpose, for the Art of directing & searching other more secret matters in Astronomy, and is thereof called the circle of Position, which passeth at al times by the former sections of the meridian and Horizone, and by the Center of the star, or of any other purposed point in heauen, like to the foresaid cyrcles, whether that star be aboue the earth, or vnder the earth. That this may clearly appeare, marke and consider this figure here expressed, where the letter c. representeth the top pointe, d. the Northerly Pole, e. the opposite
pole

of the Circles.

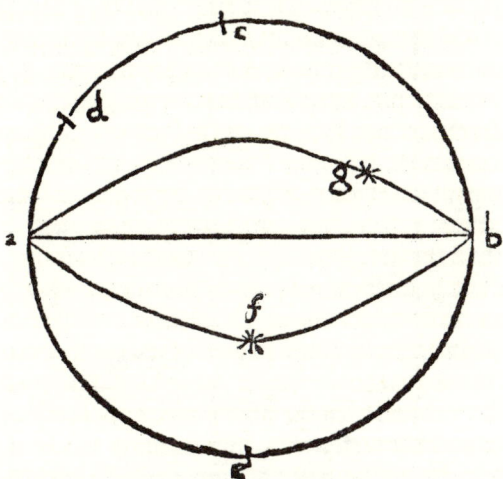

pole, a g b f. the cyrcle of the position paſſing by the ſections of the horizon and meridian, b c d e. the meridian a b: the Horizone, g f. the Centers of the ſtars (of which the one is in g. aboue the earth, and the other vnder the earth in the point f.) And many other cyrcles beſides all theſe, which hetherto haue bene deſcribed, may bee inuented and imagined in the ſphere, for the neceſſity of the workings.

The difinitions, names, and offices of the foure leſſer Circles.

He Parallels are leſſer cyrcles, which from either of the greater circles drawn thwartly on the ſphere, doe equally diſ- and bee diſtant from the Equatoure or Zodiacke toward their poles: ſo y̌ they doe not deuide the Sphere into equall halfe Spheres, but into vnequall portions

N iiii.

ons. For seeing the sphere from the middle streacheth or draweth by litle and litle straighter and narower toward the furthest and highest toppes: euen so must the parallels which are distant from the middle and greatest, and that by equall spaces on each side agreeing, drawe of necessity narrower, and so much the narower, as they nearer approach vnto the poles. As writeth Theodosius in the sirte proposition of his first Booke of the sphere. And the same Author in the 14. proposition of his first Book of the sphere, and in the sirt, of his second Booke writeth, that all the parallels haue the same poles agreeing with the greater cyrcles vnto which the parallels are.

And certaine of the Paralels are applied vnto the plain of the Equatoure, others vnto the plaine of the eccliptick. These doe as well the fixed starres, as the planets placed without the eccliptike, and drawne about the Exe-trée stretched vp the poles of the eccliptike and Center of the worlde discribe: yet do all their centers consist in the Exe-trée of the Zodiack, and the middle cyrcle of them, and the greatest is the eccliptike. These also doe the same stars, and the verticiall or toppe points of each places, or any other applied vnto the plaine of the equatour, drawne as it were by the first mouer about the Exe-trée and poles of the world define. And the Centers of these be in the Exe-trée of the worlde or equatoure, but the middle and greatest of these, is the equatour.

It is manifest by that afore taught, that the sun in euery day doth gaine toward the East (against the dayly motion) one degrée of the Zodiack: and of this hapneth, that be in each day through the thwartnesse of the Zodiack describeth a certaine newe cyrcle in heauen, and in the nexte day another, and so forth by order, as the like may be compared by a small corde, winded close about a Nun or top, beginning from the foote vpward, euen so the sun beginning to turne againe at the first degrée of Capricorne, doth

euery

of the Circles. 233

euery day after change a new Parallel, vntill hee become backe vnto the first degree of Cancer, and by and by after returned from Cancer, he in the like order goeth vnto the Capricorne: so that in the next day following, the Sun riseth not with the same Parallell aboue the Horizone that hee did in the morning before, nor shall not run the nexte morrow in that Parallel that he did in this day. And each of these Parallelles (euen as the greater cyrcles) containe 360. degrees, which bee so much lesser, then the degrees of the greater cyrcles, and occupy or comprehend somuch the lesser space in heauen, as answereth to the vpper face of the earth, as by how much the more frō the compasse and largenesse of the greatest cyrcle they lacke, by reason of the distance. And although they yeld and be lesse in the quantity, yet vnto the degrees of the greatest cyrcles be they agreeable and like, as (writeth Theodosius) in the 14. proposition of his second booke of the sphere.

These lesser cyrcles, do offer and teach sundry vtilities. First the Parallels, of which on this side and beyond the Equatour, are 182, that the sun yearly by his dayly motion describeth: and doe expresse the causes of the continuall equallity of the daies in the right Sphere, and of the onequalnesse in the thwart or bowing sphere, and where the day spaces are encreased and lengthened, there the night spaces be lessened and decreased: and being otherwise they shew the contrary.

In the second, the Parallels (which the verticial points forme) when they expresse the boundes of the latitudes of places, then are they standing vnder, by which their longitudes or distances from the West are accompted.

In the third, the Parallels (which either the Planets or the fixed stars describe) referred vnto the Equatour, do expresse the boundes of their drawings or motions from the equatour. The others or rest, which applied vnto the eccllipticke described, doe shew the bounds of the latitudes:

and

and that for how long time they tarry aboue the earth, or otherwise hid within the earth, and vnder the Horizone, doth either shew.

In the fourth, the greatest and chiefest vtilities of the Parallels are, that which on the habitable earth the practisioners seuer by such distances, as by how much ỹ greatest artificiall daies are by a quarter of an houre longer increased and extended. For they distinguish the habitable earth (and that by obseruation) into certain necessary spaces, and doe indicate the regular increasings of the daies, and what is common to each dwelling vnder those parallels, in asmuch as the quantities, the increasings and deminishings of the dayes and nightes, the risings and settings of the stars, the Noonstede shadowes, and the nature of the Winter and Summer but those which are contrary, as that there is a difference & diuersity of the dwelling places being vnder diuers Parallelles, they indœde bee necessary vnto the distribution and description of the clymate.

Although the number of these cyrcles bee so infinite, as is the infinite variety of the stars and verticall points: yet are there foure vsually rehearsed in these Elements or introduction, that be especially noted and described by peculiar names: and for the same cause (as seemeth to mee) in that they deuide the whole Globe of heauen and earth into fiue Zones, and these applied vnto the plaine or flat of the equatour.

    The tropicke of Cancer, or summer tropicke.
    The tropicke of Capricorne, or winter tropicke.
    The articke or Northerly Pole.
    The antarticke, or Southerly Pole.

Which

Which Circles are called the Tropickes.

The Sun (according to the former words) through the motion of the first mouer is in 24. houres, drawn once about: and for that hee is caried in the thwart Cyrcle, and in the same by his proper motion chāgeth dayly vnto other places of the Zodiacke, it must nedes ensue, that he describeth in each day a new parallell. And those doeth the sun repeat in the partes of the Zodiack, which be equidistant from the solsticiall points; in such wise, that they be in the whole 182. cyrcles. And these do they call the cyrcles of the natural daies, of which the vttermost and furthest that include the suns way, are named the Tropicks, which is (in English) the sun bouds, in that the sunne neuer passeth them, neither toward the North nor toward the South: but after his touching of each, he returneth againe. The one of these called the tropicke of Cancer, and the other the tropicke of Capricorne.

Why these are called the Tropickes.

They are named the Tropicks, of the Greeke word *Tropikoi*, which is in English, the turnings againe; in that when the Sun is digressed from the Equatoure and come vnto those, hee turneth backe againe. Also the Tropicke cyrcles touch the Zodiack, at the beginnings of Cancer and Capricorne, of which the one is calis called the Tropicke of Cancer, and the other of Capricorne, the one being Northerly, and the other Southerly. And as to our dwelling, the one is called the summer Circle, and the other the Winter. So that when the sun toucheth any of these, he turneth againe, and is carried toward the other. As by this example further appeareth, where all that season and time (from the twelfth day of December vnto the eleuenth day of June) a manne may perceiue the Sunne euery day arising higher and higher; and when he is at the highest ouer our heades, that day doth he by his course describe the summer Tropicke: from which againe turning, the sunne euery day after draweth lower and lower from our verticall pointe, vntill he be come againe vnto the lowest. In which twelfe day of December (not going any further toward the South, but being come vnto the beginning of Capricorne) he describeth the winter Tropicke.

The Tropicke of Cancer is a lesser Circle, which the sunne describeth at the entring into the beginning therof, and is drawne by the daily motion, whose plaine or flat passeth not by the center of the earth: and it is one of the naturall Circles which is outermost, described of the sun toward the North, and drawne by the beginning of Cancer. And it hath also his name of the standing, in that the same is the bound of the sunnes iourney or course toward the North, and the nighest comming vnto vs: vnto which being brought, he turneth backe, and directeth his course into the South; of which that place is called *Trope*. It is continually distant from the Equatour, by the quantity of the suns greatest declination, which at this day is of 23. degrees, 28. minutes, and two fifts almost: and it encloseth also the suns way, and doth besides, with the other 3. Parallels, deuide the Zones of heauen and earth. Further, this is named the cyrcle of the summer solstice, by the same reason, in that it is drawne by the pointe of the summer

of the Circles. 237

ner solstice. And the Northerly Tropicke in that it is the Northerly part of the world. And the summer cycle, for that the Sun in the summer falleth into this cycle. Also this cycle in all the Northerly tract is on this wise, that the greater part or portion is aboue the Horizone, and the lesser part (as to vs) vnder the Horizon: so that the sunne runing in that cycle, causeth the longest day of summer. And whiles the sun describeth these cycles, the dayes bee longer then the nightes. For the longest day increaseth from minute to minute, from houre to houre, and from the latitude of one degrée, vnto the latitude of 66. degrées, and 30. minutes. In which the day artificial is of 24. houres, and is thereof called a whole day. For in the latitudes following, and beyonde, hee increaseth into many whole daies.

A like definition hath Proclus, where hee writeth that the summer Tropicke, is the furthest cycle Northwarde that the sun describeth: into which when the sun is come, he then maketh his summer turne, and causeth also at that time the longest day and shortest night of the yeare: from which turning backe, he goeth againe toward the contrary coast of the world: so that of the same Proclus it is called a Tropicke (which is in English) a returning cycle. For it is euident to all men, that after the sunne beginneth to turne, he may in short time after, or at the least within 5. dayes, but especially at Noone in euery wéeke, be well perceiued to discend and go lower and lower, vntil he become vnto the Tropicke of Capricorne or the winter cycle: where he turneth againe, as you may plainly learne and vnderstand by the former description of that cycle.

The Tropicke of Capricorne is a lesser cycle, and one of the naturall cycles, which is by the like space distaunt from the Equatoure into the South, and described of the sun in the beginning of Capricorne, as being vttermost toward the South (which is the bound of the suns greatest

The second Part

departure from vs, and of his longest digression vnto the South) that he defineth and maketh. This cycle also is called the winter Solstice, and winter Tropicke; in that when the sun cōmeth into this cycle, it is presently winter: that is, the shortest day of the yeare. Also the lesser portion of this cycle is to vs aboue the horizone, and the greater beneath or vnder the Horizon. Besides the suns iourney endeth at the south, and crosseth or deuideth both the burning and temperate Southerly Zone.

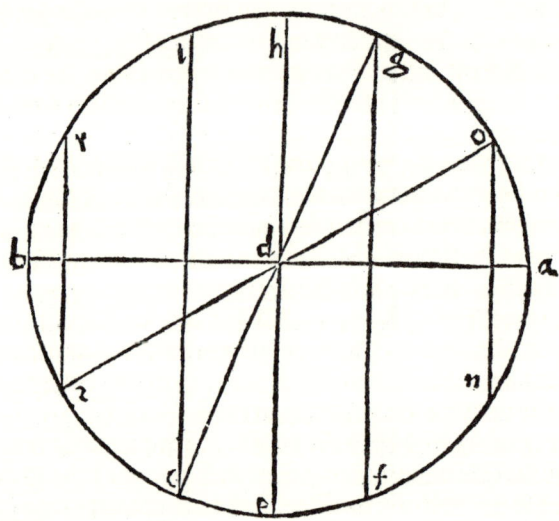

The Brumall or winter tropicke (as writeth Proclus) that is furthest toward the South, of all those which the sunne describeth by his drawing about of the first moouer: into which when the sunne falleth, he causeth his winterly returne: so that the longest night of the yeare and shortest day, is at that time procured. From which he goeth

of the Circles. 239

no further toward the South, but returneth vnto the contrary quarter of the world: and hereof this Circle is called a Tropicke, or circle of returne. Now these three are principally noted: the equatour, and the two Tropickes, for the course of the sunne. That the instructions of the two Tropicks afore spoken of, may more perfectly be vnderstoode, conceiue this Figure heere demonstrated: In which a h b c. is the meridian, a d b. the right Horizon, d. the Center of the principall a. the Northerly Pole, b. the Southerly Pole, g d c. the cyrcle of the Zodiacke, h d e. the Equatour, which here is ment to bee abatingly described, when the sun shall bee in the center of the earth, or in the true section of the Equatoure and Zodiacke, as in the letter d. from which point when the sun returnes toward the Northerly pole a. vnder the cyrcle g d c. he then describeth in each day by the motion of the first mouer each Parallels, vntill he be come in the Meridiane vnto the point g, from which hee can not further ascend toward our Zenith in the meridian. Of which in the same day, the sun describeth g f. the parallell to the equatour, which is called the Tropicke of Cancer, in that the sun beginneth from this place to approach or draw nearer to the Equatoure; vnto which when the sun shall come, hee then descendeth vnto the neather halfe sphere, in the halfe cyrcle d c. Hee being come againe vnto the point c, doth from his center by the motion of the principall or first mouer in the Poles of the world, describe the Parallell c i. that is, the tropicke of Capricorne.

The Polare cyrcles, are two of the lesser cyrcles neare to the Poles of the world, being alike equall distant to the equatour, which vpon the Poles of the equatour described are drawne by the Poles of the Zodiacke. And these are named the Polare Cyrcles, in that they bee neare to the Poles: of which, that neare to the Pole articke, is called the arcticke cyrcle, of the greater or lesser Beare drawne

in

in it, or for that this cyrcle is described about the pole articke: the other that is right againſt, is named the antarticke Cyrcle, in that it is drawne aboute the Antarticke Pole.

Or thus, the arcticke is a leſſer cyrcle, which the Northerly pole of the Zodiacke is ſo far diſtant from the Pole articke of the world, as is the ſuns greateſt declination: or as Proclus writeth, that the fore foote of the greater Beare by the dayly motion formeth. The ſame cyrcle (after the minde of the learned) is diſtant from the equatour 66. degrées, and 30. minutes almoſt. To whome this altitude is higher by 23. degrées, and almoſt 29. minutes. To thoſe parts of the earth is the pole arcticke extaunt in ſight, and continually appeareth. It alſo ſecludeth and parteth the untemperate Northerly Zone, from the next temperate Zone: where the Solſticiall Tropicke is made the Northerly cyrcle, and in that place under this altitude of the pole 66. degrées, and 31. minutes: there all the ſtars and images contained from the ſolſticiall Tropicke unto the Pole are ſéene: as both the Beares, the Dragon, Cepheus, Caſſiopia, Perſeus, Auriga or the Carter, whole Bootes (except from the knées downeward) the crowne Hercules (except the head and right arme) the Harpe, the Swan, the great Horſe Andromeda (except the left Cubit) the halfe of the Northerly Fiſh almoſt, Deltoton, a part of the backe of Taurus, the necke and Northerly Horne, a great parte of Gemini, and the head and necke of Leo.

And not unlike to the former, doth Proclus deſcribe them: where hee writeth, that the Northerly cyrcle is the ſame, which of al thoſe that to us continually be ſéen or appeare, is for trueth the greateſt; and that alſo toucheth the Horizone at one onely point, being wholy deſcribed aboue the earth. And the ſtars that are incloſed within this cyrcle, do neither riſe nor ſet, but are continually ſéen all the night drawn about the Pole.

The

of the Circles.

The South or antarticke cyrcle, is thus defined of him, that the same is equal and equidistant to the Northerly or articke cyrcle, and toucheth the Horizon at one point. The whole of this cyrcle is hidden vnder our Horizone, so that all the stars placed and drawn in it, abide euer out of sight to vs.

The like description that the antarticke Parallell is a lesser Cyrcle, which the Southerly Pole of the Zodiacke drawth about as it were by the dayly motion, doeth describe about the Southerly toppe of the world, and is by a like space distaunt from the Equatoure and the antarticke pole of the world, as the articke is from his opposite. And doth seperate or deuide the vntemperate Southerly Zone from the next temperate Zone.

Further it is manifest, that the distance of the Poles of the ecclipticke from the poles of the world, doe agrée with the greatest bowing or declination of the ecclipticke or the sun: In that the poles from their cyrcles, bee alwaies distant a quarter of the cyrcle, and the colure of the solstices, is here taken for that which comprehendeth either Pole. And when the quarters standing betwéene the poles, and the cyrcles of the poles, be in themselues, or betwéene one the other equall, as the arke of the same cyrcle, then the middle arke common to both, which (as exempted) goeth betwéene the poles of the world and the ecclipticke, and so parteth and leaueth them equall. For the one halfe of the other equall arks, is from the poles of the ecclipticke vnto the poles of the world, and the other, is from the furthest point of the ecclipticke vnto the equatour. By which it appeareth, that so much is the distance of the poles of the Eclipticke from the poles of the worlde, as is the suns greatest declination, being 23. degrées, and 28. minutes, and two fiftes almost. Or thus, that the pole of the Zodiacke is far distant from the pole of the world, as is the greatest declination af the sun from the Equinoctiall cyrcle: and by

R j.   the

the equidistance also on each side of the arctick cyrcle from the Pole of the world, that that part of the Colure comprehended betwæne the first point of Cancer and the articke cyrcle, is almost double so much vnto the greatest declination of the sun.

And if cyrcumspectly you consider the maner of the motions, you shall readily perceiue that those cyrcles which euer more be of like largenesse, increase and decrease togither with the twoe Tropicke cyrcles, accoording to the increase or decrease of the suns declination. As appeareth by the letter n. in the foresaid figure, that representeth the Northerly pole of the eccliptike or Zodiack, moued from the letter n. into o. by the motion of the first moouer, and returning againe into the point n, shall be moued the cyrcle describing n o. being distant from the Northerly pole a. as much as is the suns greatest declinaton h g. as hereafter by demonstration shall plainer appeare. And this cyrcle named the arcticke, in that it is described by the arcticke of the Zodiacke. The like is described from the point r. being the pole antarcticke, by the motion from r. vnto s. and returning againe vnto r. so that the antarcticke cyrcle r s. is equall to his opposite, and equidistaunt to the Equatoure.

This probation, that the distaunce of the Poles of the worlde and Zodiacke, is equall to the suns greatest declination, doth require before hand, these three propositions. The first that the quarters of each cyrcle any where taken be in themselues or betwæne one another equall. The second, that the poles by a quarter; that is, by 90. degrees, be distant from their proper cyrcle. The third, that the equals deducted from their equalles, then doe the equalles rest.

As for example, if you borowe two fourthes in one and the same Colure cyrcle, that is the Solsticiall of the same parte, where it passeth by the beginning of Capricorne,

and

of the Circles.

and is the like from the pole of the worlde vnto the Equinoctiall, and that other, is that which is from the Pole of the Zodiack vnto the Zodiacke or ecclipticke: and of this I thus reason, that when the equals be deducted or abated from the equals, the remainer shall be equall. Therefore are the foresaid quarters equal, in that they be in the same cyrcle, and that from either is the equall or common arke deducted; that is, the same which is contained betweene the Equinoctiall and the pole of the Zodiacke, which arke doeth containe 66. degrees, and 31. minutes almost. So that the arks, resting or remaining of these quarters be equall; that is, the distance of the poles of the Zodiack, and the Equinoctiall, is equal to the suns greatest declination. For if 66. degrees, and 31. minutes bee deducted from either quarter, the remainer then shalbe 23 degrees, and 31 minutes: which is the distance betwæn the foresaid poles and the greatest declination of the sun.

This other example demonstrateth, that the suns greatest declination and the distance of the poles of the zodiack or eccliptcke from the poles of the world, is equall and of like largenes, and that what soeuer hapneth to the distances of the said Poles. For as this increaseth or decreaseth, the like doth that decrease or increase. Of this it is manifest, that the two foresaide articke cyrcles, is nowe in our time lesser through the decreasing of the suns greatest declination, and that the Tropickes are greater then they were in Ptholomies time.

The example here followeth, where i f. representeth the Exe trée of the world, e a k. the Exe-trée of the Zodiacke, c l m n. the Meridian or Colure of the Solsticis, c m. the Equatour, b l. the Tropicke of Cancer, d n. the Tropicke of Capricorne, I. the Pole articke, F. the Antarticke and opposite pole to it, k. the pole of the Zodiacke, and e. his opposite, b c. the greatest declination of the Sunne, which in our time the practisioners haue founde to be of 23. de-

grées,

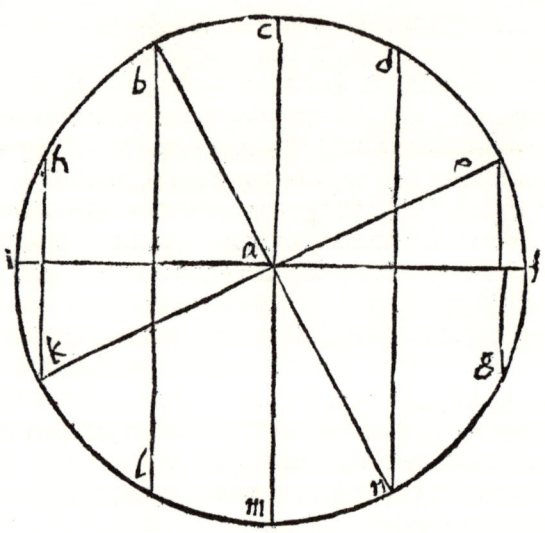

grées, 28. minutes, and 30. secondes: which in Ptholomies time was of 23. degrées, 51. minutes, and 20. secondes. The letters c g. and h k. be the two articke cyrcles. The arke b c. is equall to the arke i k. in such wise, that it may plainly appeare that b h k. and c i k. be the two quarters of one and the same cyrcle, f c k m. All the quarters of the cyrcle bee a like equall one to the other, c b i. and b i k. bee the two quarters, in one and the same solsticial Colure cyrcle. So that c b i. is equal to that b i k. and the ark b i. is to both the quarters equall or common. If then the equalles bee from the equals deducted, the remainer shall be equall, as aboue taught. For from the saide equall quarters, c b i. & b i k. deduct the equall, that is b h i. or the common. And the remaining arks, of the said equall quarters, shall be in themselues equall, that is, c b. shall bee equall to the arke

i k.

ik. as agræth to be wrought.

The offices or vtilities of the foure lesser Cyrcles.

1. The office of the Solsticiall Tropicke: (after the Grækes) is to define the longest summer day: and the winter Tropicke, to determine the shortest winter day and longest night. For Proclus to finde the longest day, did deuide the summer solstice into 8. equall partes; of which so deuided, he affirmed thrée parts to be continually hidde vnder the horizon, and fiue aboue. The truth of which is known, if the Sphere bee rectified for the latitude of 41. degrées, whereby this diuisiō the longest day containeth 15 hours and the night but 9. houres.

2. Many and notable offices doe the Tropicke cyrcles offer, as well vnto the composition of dialles, as vnto the preparing many other Instruments in Astronomie.

3. They declare the places of the Ecclipticke, in which the suns solstices are caused: whereof the longest or shortest daies by them are knowne. Or thus, they declare in euery standing of the sphere, the longest and shortest day, with their quantity.

4. They include the Suns way, in that they bee as the bounds including the Region in heauen, in which the sun is continually moued.

5. They declare the suns greatest declination, as afore hath bene often taught.

6. They seperate in heauen, the burning Zone, from the two temperate Zones.

But of the Polare cyrcles, these be the chiefest and especialest vtilities.

R iij. 1 They

1 They iudicate or shewe the Poles of the Zodiacke, and howe farre they bee distaunt from the Poles of the world.

2 They inclose those stars which euer appeare aboue our Horizone, and those in like maner right against being alwaies hid vnto vs. But for that euery seuerall Climate (hath disagréeing from other Climates these cyrcles) their distance therefore cannot bee certaine from the other Parallell cyrcles, sauing for one Region certaine, as neither their quantities, nor their order. For in that place, where the altitude of the pole is lesser then 66. degrées and a half, these cyrcles there are lesser then the Tropicks, and in order are betwéene them and the poles, and is from the pole continually distant by so many degrées, as the pole in that country is raised aboue the Horizon. So that in the same place, the Pole raised more then 66. degrées and a halfe. The Tropicke then is aboue the horizone, as the like may be vnderstoode by that place called Wardehouse. So that in the same Climate, the arcticke cyrcle is greater then the Tropicke of Cancer, as witnesseth the learned Stoeflerus, Iustingensis.

3 They distinguish (after the mind of the Grǽks) the cold Zones, from the temperate. Which Ferio denieth, affirming that the arctick and antarcticke cyrcles, kéeping no vniformitie to all countries and béeing vncertaine and variable boundes, can limit any certaine place. For the temperate Zones are places certaine, the arcticke and antarcticke cyrcles bee changeable limits, therefore cannot they be as bounds of the temperate Zones: yet dooth hee better allowe and agrée vnto that, that the Tropickes bee bounds of the temperate zones. So that changeable limits (by this argument) cannot be appointed as bounds to vnchangeable places.

4 They deuide togither with the Tropicks, all heauen into fiue parts or Regions, which they call zones.

The

of the Circles.   247

The descriptions, names, qualities, and vtilities of the Zones.

THe foure lesser cyrcles called Parallels (that were afore described, doe deuide the whole heauen towarde the Poles into fiue spaces: which that heauen might bee compassed aboute with these larger swathes, the astronomers of the same called them Zones, or otherwise of the Latines Gerdils. The Cosmographers by the same imagination applied, doe also dispose and distribute the whole Globe of the earth into fiue roomes or spaces, lying directly vnder, and agreeable in proportion to them in heauen.

Wherefore a zone (after the minde of the Grækes) is a portion, tract, or space of heauen, or earth, betwéene the two Parallels or lesser cyrcles, being nighest equidistant, or contained betwéene the roome equidistaunt and Pole of the world, and gyrdeth or compasseth as it were the heauen or earth. Or thus, a zone is a space of earth like to the two Parallels or lesser cyrcles aboue, which the astronomers imagine to run on the vpper face of the sphere. And as the whole portion included by the two Tropicks called the burning zone, doth compasse heauen as a gyrdle: euen so imagine the roome of the earth, lying right vnder the Tropicks.

The zones haue sundry names, for of the Grækes they be called zóne, and of the Latines by a borowed word Zona, as may appeare by Iulius Firmicus, Macrobius, Virgilius, Ouide, and other Latines. That heauen or earth is imagined to bee gyrded about with these. Martianus na-

R iij.   meth

meth them swathes, Tully and Macrobius nameth them by the like reason gyrdles. Ouide nameth them plagues; that is, roomes or spaces.

And how many zones they bee, may easily appeare, in that the astrologians, Geographers, Phisitions and Poets, do deuide as well the heauen as earth into fiue roomes or spaces, by the foure Parallels or lesser cyrcles: of which there bee two maner of zones: the celestiall and the earthlie.

The celestiall, are the cause of the earthly, in that the earthly lie directly vnder them. And of the zones, the celestiall bee they which the astronomers by imagination describe and distribute in the hollow of heauen: the earthly, be they which lie perpendicularly vnder. And both also be temperate, and vntemperate zones.

The celestiall zones, in that they haue nothing of the elementary qualities, therefore doe they not by heat burne and scorch, nor by cold make stiffe: nor cause a temperate mixture of qualities or temperatnesse, yet are they noted and descerned by the names of the qualities; as the earthly zones, which being the author of the sun, and fountaine both of light and heate, and running continually in the middle zone of heauen is diuersly felt, according to the maner of the distance.

Or thus, there are no qualities formally attributed to the celestiall zones, but to them onely vertually, which is on this wise to be vnderstode, as that the celestiall zones of themselues be neither cold, hot, nor temperate, but are so called through the suns declination from the equatour, as well into the North, as into the South quarter of the world: In the which declination, is the like matter felte, as well in the suns right sending downe of beames, as in the thwart proiection of thē on the vpper face of the earth, which diuersly changeth the heat &c.

The scorching or vntemperate middle Zone (which
through

of the Circles. 249

through the heat and burning beames, the sun there causeth, when he is ouer the head or in the Nonestæd place) is contained betwæne the boundes of the sunnes iourney which the two Tropicks make, and includeth 47. degrees of heauen. For the two Tropicks are on either side the equatoure, so that it vseth the middle rome in the burning zone, from which the sun towarde the North and South, neuer declineth aboue 23 degrees, and 29. minutes. By which appeareth, that it is there as hot in the middle of winter, as it is in Spaine in the middle of summer: and therefore not disagræing to that which the auncient Cosmographers wrote, that the countries lying vnder this space, or rather vnder the equatour, is vnhabited through the burning heate: and of them for this cause, named the burning or scorching zone. But of later yeares it is found contrary, in that at Molucca, Good-hope, Calicute, and Samatra, rich drugges, and other fine spices haue beene there gotten by the Spaniards and Portingals, and yærly haunted by them, as at this day the same is throughly known to many: which also confesse that the places vnder the Equinoctiall, and the rich City Calecute, being by the sea coast of Inde, standing betwæne the equatour and our Tropicke of Cancer, and vnto the other Tropicke South vnder the burning zone, that the places is habitable and peopled, although very cumbersome with extremity of heat. Also that space on earth containeth 685. Germaine miles, or 23500. furlongs.

Ptholomie and Auicen affirme, that the places betwæn the equatour and summer Tropicke is habitable, and that many Cities bee there, although the sunne in those places through his direct beames and especially vnder the equatour) doth by the ouer much heat and continual heat, burn and mightily scorch. The like doe sundry others affirme, which write, that those places is conuenient for the life of creatures, in that vnder the equatour there bee many waters

ters, which although resolued and run through the heate, yet doe they breath and send vpward colde vapors, which the sun continually maintaineth in drawing vp through his vehement heat, and sending down mighty showers of raine: which vapors in the night (through the suns furthest distance vnder the earth, and especially at midnight) cause a mighty cold and chilling ayre: which the sun after his rising, vntill he be somewhat ascended aboue the earth cannot sodainly ouercome and put away that cold impression of the ayre. So that the people there inhabiting, bee monstrous of forme, and haue rude wits, wondrous wild and terible conditions, like to wilde and furious beasts.

The countries which lie vnder the Southerly Parallels, as those which are described by the Equinoctiall line, vnto the summer Tropicke, where the sun is drawne and runneth ouer the tops of them: there through the aboundance of vapors, rayne, and night colde, is the suns heate repressed, mitigated, and dulled; so that the heades of the Ethiopians or Moores be litle, hauing but litle and withered braines, their bodies short, hauing thicke crisped haire on their heades, grosse and dull of senses, blacke scorched or burned bodies, withred or wrinckled faces, croked of stature, being in a maner hot by nature, and cruell condicions, through the mightinesse of heat in those places. And the constitution also of the ayre is there such, that al liuing and cresent things on that earth, are found and known to agrée with them. Further it is to be noted and vnderstood, that any there trauailing from the Northerly places, the further they goe towarde the South, somuch the stronger heat or burning they shalbe annoyed with.

The two temperate zones be next adioining to the burning zone, the one on the Northerly, and the other on the Southerly side of it. And the beginnings of either bee the hotter, the ends colder, the middle of them exquisitly temperate: in the other parts doth the heat either so much the

more

of the Circles. 251

more excǽde, or the bitter colde ouercommeth and ruleth, as howe much the nearer they approach or come vnto the burning Zone, or otherwise vnto either of the extreame Zones, which continually cause a bitter and an extreame colde.

The cause of this diuersity, is through the suns beames, for the sun continually moouing in the middle iourney of heauen (described betwéene the two Tropicks) and digressing or going beyond the prefixed bounds of nature, doeth not shew his beames vnto diuers parts of the earth in one manner, but vnto the places right vnder, and in the burning zone the tractes or countries contained vnder them, doth he send downe right beames, which stretcheth to the vpper face of the earth at right angles. And vnto the countries of either temperate zone, doeth the sun send downe thwart or slope beames. And vnto the places vnder either cold zone, doth he streach long beames on the plaine of the earth, euen the like as being neare to the Horizone, which neither reach vnto the vpper face of the earth, nor cause angles, but kéep an equall distance vnto it, do streach forth infinitely.

But those beames of the sun doe neither giue light, nor heat, but turne backeward: in that the property of the reflexion which of the beame against a solider resistance, prohibiting or letting the penetration, is a certaine repercussion and reuerberation) that increaseth and doubleth the force of the direct beame, and by the reflexed beame to it adioyned, or at the least by his vertue applied and communicated.

Séeing this reflexion is the especiallest cause of the heat and that the angles of the reflexions falling doe continually make or be equal in the angles: for that cause do they much vnlike increase the force of the directe beames, and their effectes doe notably varie. So that in the burning zone, the reflexion stretcheth vnto right angles, séeing the
straight

straight or right beames are caried & led into themselues, in such sort that as direct and resteren, they meete and bee mixed, and in this, doubling as it were the vertue and force of the direct beames, is on such wise increased, that it kindleth, burneth, and consumeth.

And in either temperate zone, is the reflerion caused at right angles in that the sun beames doe thwartly reach to the vpper face of the earth, and are turned and extended backward vnto thwart angles, which how much the neerer and liker they bee to the right, so much the nearer doe they ioyne either beames togither: by which they proceed and come into the nearer parts of the burning zone. But so much the blunter as they streach, so much the longer do they seperate either beames, as howe much the more they are extended vnto the extreame or outmost bounds. And for this cause doe they more heat then the fore parts of the temperate zone, whose heate is a litle gentler or milder then the heate of the burning zone, and the beames a litle further of: whose colde notwithstanding differeth somewhat from the extreame or outmost vntemperate zones.

And those which streach and fall into the middle region of either temperate zone, doe cause a meane betweene the right and very sharp angles, and yet not directly matched or ioyned, nor doe they by so neare a space communicate their vertue, as in the beginning of it, neither by so large a distance as in the end, but in the middle in a maner: So that they cause and increase a temperate heate in the same zone.

But in the extreame or colde Zones, is no reflerion of beames caused, for those beames equally distant from the earth are streached forth infinitely: and for that cause doe those neither giue light nor moue or procure heat, neither doe those zones at any time warme, either perfectly cleare, or appeare bright: but that they continually be foggy, misty, darke, and bitter or extreame cold, through the continuall

of the Circles. 253

nuall mists, which more and more increase, especially toward the northerly pole. And yet many affirme, a reasonable dwelling in those places, yea and vnder the Northerly pole, but far colder and bitterer dwelling, through the far being from the way of the sun, and beholding of the comfortabler starres. For the Sunne through his ouer far distance, cannot by his presence aboue the earth comfort and heate.

This now is the perfect cause of the diuers and vniuersall constitutions of the ayre and chiefe qualities in each zones: so that of the particular constitutions be other causes. But to returne vnto the temperate Zones, the latitude of either temperate Zone is of 43. degrees almost, of Germaine miles 645. and of furlongs 21500. So that the Boreall or Northerly zone beginning from the Tropick of Cancer, endeth at the arcticke cycle, or at the degree of latitude 66. and 31. minutes. And the Southerly from the Tropicke of Capricorne, is extended or reacheth vnto the antarcticke cycle, or the degree of the Southerly latitude 66. and 32. minutes.

The vntemperate cold zones that reach frō either temperate vnto the poles of the worlde, doe mooue continuall cold and frosts. So that the beames of the sun, although they pearse and enter through, yet seeing they extende not backward, nor through the reflexion or streaching backeward be strengthned and sharpned, therefore can they not so heate, that by the thawing they dissolue the earth and yse, nor put away or voyde the mist. Now the vntemperate Northerly zone, beginning from the 66. degree and 31. minutes of the Northerly latitude, endeth at the Pole arcticke: and the vntemperate southerly zone, begun from the same bounde of the Southerly latitude, extendeth and endeth at the pole antarcticke.

Those people which dwell vnder the burning zone, bee named of the Greekes *Amphiskioi Amphiscij*, in that the

Nnne

The second Part

Noone shadowes, at diuers times of the yeare, goe or be cast to them twoe waies, as toward the South or North. And twise also in the year runneth the sun right ouer their heades (as is demonstrated in the second Theorme of Euclide) so that at Noone it commeth to passe, that they haue almost no shadow: for the sun being direct or in right line ouer their heades at Noone, hee then sendeth downe right Beames, which are cast or streached to the plaine of the earth at right angles: so that their shadowe falleth and is right vnder the feete, and not on any side of them. So that the sun in any other time of the yeare beeing without the verticall pointes, the shadowes at Noone are one whiles cast into the South, and another whiles into the North vnto them: euen as the sun digressing from their toppes or Noonsted is either caried into the North, or otherwise declineth into the South.

This sorte of people which bee vnder either temperate zone, are called of the Græke Cosmographers *Eteroskioi, Heteroscij,* in that they haue a single shadowe. For with them the Noone shadowes continually run or goe toward one quarter onely. So that to them dwelling Northward the Noon shadow streacheth towarth the articke or Northerly quarter. By which it appeareth that the sun neuer ascendeth ouer their heades, but continually casteth or streacheth his beames thwartly into those contries, which alwaies forme their thwart angles with the plaine of the earth, or els fall a slope vpon the earth.

Those people which possesse, and dwel vnder either vntemperate or cold zones, are named of the Græke writers *Periskioi, Periscij*: for that their shadowes in one artificiall day are caried and run rounde, as it were about them on the plaine of the earth: so that the sun vnto those places casteth or sendeth not straight, thwart, or sloape, but long beames running on the horizone, which as they streach along infinitly; euen so the shadowes going and lying on the

of the Circles.

the flat of the earth, and extended along, doe increase infinitly. And these zones also vnder the poles, extend to that proper place, where the Tropicke cycles, and the Arctick cycles be all one. Strabo likewise writeth, that the colde zone reacheth to that place, where the Tropick is the arcticke cycle; that is, where this first Zone endeth, and the temperate beginneth: the Pole beeing 66. degrees, and a halfe aboue the horizon: so that this pole must be from the toppe of their heads in that place 23. degrees, and a halfe. Further, these people that haue their shadowes running rounde about them, dwell within the Polare cycles: In that all people whose Zenith is within 23. degrees and a halfe of any of both the Poles, haue their shadowes compassing aboute them, but these people (as afore written) dwelling nearer vnder the Pole, the longer is their day, and by that reason doe the shadowes run the oftner about them. For where the day is of 24. houres long, the shadow doth run but once about, yet where the day is of halfe a yeare long, the shadowes doe run 183. times about.

Here conceiue that there be fiue zones on earth, answering to the fiue celestiall zones, both in the heat, temperatnesse, and cold: which for a plainer vnderstanding, vse this figure here following demonstrated. Where the Orb or cycle described on the plaine of the earth, is distributed by the two vnknown diameters into foure equall partes: as to the outward points of the one diameter, note the letters a b. To the points of the other diameter, adde the letters c d. The letter c. the Northerly Pole, and the letter d. representing the Southerly pole. The arke of the Orbe a c. deuide after the common maner into 90. parts or degrees, the number (as the vse is) noted by 5. 10. 15. 20. 25. &c. And beginning to recken at the letter a. in ascending by the number 5. vnto the letter c. beeing the Northerly pole.

Further from the a. toward c. which demonstrateth the
tins

256   *The second Part*

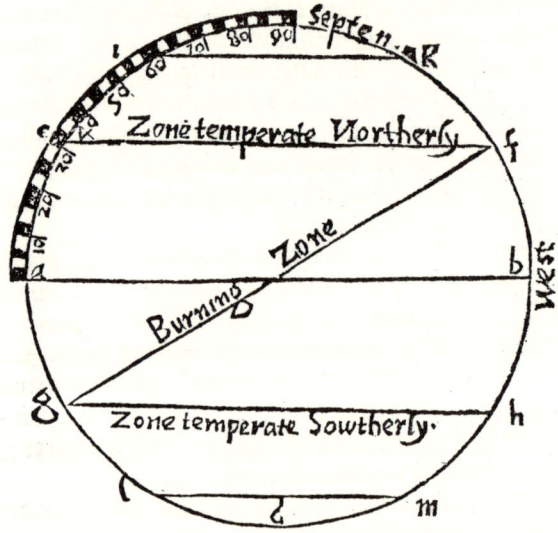

suns greatest declination, that containeth in our time 23. degrees, and 28. scruples, to the ende of the supputation adde the letter e. And by the office or vse of the compasse, comprehend the arke a e. And by the foote also staying on a. reach with the other foot of the compasse opened toward d. the Southerly Pole, to which adde the letter g. After set (the compasse not yet varied) one foote on the pointe b. ⁋ with the other make a note or pricke toward c. to which ioyne f. and drawing the compasse againe make another point toward d. which betweene note the letter h. Againe from the letter e. draw a straight line vnto f. which represenenteth the Tropicke of Cancer. And from g. likewise drawe a line vnta h. representing the Tropicke of Capricorne.

Againe set one foote of the compasse vnchaunged on the letter

of the Circles.

letter c. representing the Northerly Pole, and with the other opened, make a point of the one side, and after on the other side, in drawing a line besides from point to pointe, and the ends of that line note with the letters i k. that declare the arcticke cyrcle. And likewise set one foote of the compasse on the note d. iudicating the Southerly pole, and after the marking with points on either side, draw a right line; at the ends of which, note these letters l m. that represent the antarcticke cyrcle.

These so finished, you shall then see on that plaine or flat the fiue earthly zones rightly described. For the space here represented of the earth by the Tropicke lines e f. and g h. included, doth demonstrate the burning zone. In that the suns heat by his direct beams ouer it, doth continually strongly heat and burne that space of the earth: wherefore you may rightly draw a straight line from the letter g. vnto f. representing the suns iourney.

And the space of the earth included of the line i k. the arcticke cyrcle, and the arke answering to the Northerly line i k, doth iudiacate the cold and frozen zone Northerly. And that other Region or space contained right against, representing the antarcticke cyrcle, doth demonstrate the colde Southerly zone. And the tract or space of the earth included within the lines i k. and E F. doth indicate our temperate zone Northward, and that reasonably habitable: and the other portion of the earth contained within the lines g h. and l m. doth manifestly shew the temperate southerly zone.

Where Ptholomie and other auncient Cosmographers write, that the burning zone is vnhabited, or as a desart, Aristotle, Plinie, and Iohn de sacro bosco (in his treatise of the Sphere) write the contrary: Besides these, it is well knowne at this day, yea by experience vnderstood of those that haue yearely gone and come from the countries lying vnder that zone; that is, betweene the two Tropicks to be inhabited. Further this burnt zone is inhabited and well replenished with people that there dwell, as the same is throughly knowen to many that haue passed to and fro the Indies: so that it may euidently appeare, that the heate there is not extreame, nor so distemperate, seeing the time of the heate that they suffer, continueth not long, nor the heat sharply worketh or causeth his vttermost effect. For the sun but a small time tarieth aboue the Horizone in the burnt Region or Zone (as certain astronomers write) as the space of twelue houres onely: so that the heat there is much qualified and suppressed, through the colde rising in the night time; whereof it is manifest, that he causeth not his extreame hotnesse there, although hee streacheth his beames perpendicularly on the earth. Therefore may many maruaile, that sundry ancient men affirme these parts to be vnhabitable, seeing they knew of Arabia, Fœlix, Aethiopia, Taprobana, and diuers other contries lying vnder the burnt zone: yea besides these, are Guinea, Calicute, Muluca and Gatigara, well knowne to lie or bee vnder the burning Zone: and many of the people in those countries liue long: and the same Region also is inhabited and replenished well with people. A like affirmation hath Albertus and Auicen (as afore written) that the middle zone is habitable, for they agree cōtrary to the old writers, that in the same Region of the world, which the auncient Cosmographers named to bee the burnt Zons, that it is a far temperater dwelling, than vnder the Tropicks it can bee in any wise. And that people dwell vnder the Tropickes, the ancient neuer doubted. Wherfore if so resonable dwelling be vnder the Tropicks, it cannot be otherwise (as affirmeth Petrus de apono) that vnder the Equatoure, (notwithstanding the sunnes sharpe heate) but that men may dwell there for all the vntemperatnesse of heate. To vse briefe, al the writers of later yeares agree, that the middle zone is not onely habitable, but found and knowen by many

of the Circles. 259

ny reasons, and by experience, that the same is most temperate, and the earth vnder it rich, both of golde and rich drugges, and reasonable well furnished of all things needfull for mans life. So that in the same middle Region of the earth vnder the Equatour, it appeareth, that through the coldnesse of the night, it doth there temper sufficiently the burning heat of the day. Besides these, after the mind of Hiero. Cardane, in that Saturne, Mercurie, and the moon (which properly are cold and moyst planets) haue a great force in the Regions vnder this zone, but especially the moon, that worketh her most force there in the night time, more then the other twoe: and of this cause more temperatnesse in the day time. Besides these, it is well known that those people haue two summers, and two winters in the yeare. For in the yeare of our Lord 1530. at the will and charge of Charles the fift Emperor, a parte of America westward was discouered, where Peru among the rest, was found richest both of Gold and other rich things and costly drugges, which is situated in longitude, of 290. degrees from the West toward the East, and is distant 5. degrees from the Equatoure toward the South. But what substance of Gold and other rich things hath yearely bene brought from this yle, needeth not here any further rehersall. And the like is to be considered and noted of the other two zones, contained betwéene the Polare cyrcles, and Poles of the worlde. Although Albertus Mag. denieth, a commodious dwelling for men in those places, and confirmeth the same by probable reasons, yet experience reclaimeth and denieth those opinions of his, and other ancient writers. In that it is well knowne that Gothland, Norway, Russia, Lapeland, Groueland, and diuers other countries towarde the North pole, is inhabited and well peopled. And Galeottus Naruiensis proueth, that men dwell vnder the North pole, affirming the same not to bee true, that the cause of the cold there is onely the far distaunce of

S ij. the

the sun, as not of the heate by nearenesse of his comming. In that the sun by reason of the signe in which he is, either increaseth or diminisheth them with vs. Besides he affirmeth, that the colde is not so dispersed rounde about, as that it compasseth rounde after the forme of a cyrcle, nor that the heate in like sort doeth run round about the whole body of the earth. Further Cardane writeth, that vnder the poles, there is no such coldnes as some suppose, in that the Moone, Venus, and Mars, haue the greatest latitudes, in respect of the sun, and the others besides. For the moon hath fiue degrees to the North, Venus and Mars exceed vnto eight degrees Northward, but Saturne which is the author of cold, scarcely performeth three degrees Northward. Besides these, the Moone more auaileth Northward and Southward neare to the poles, then the sunne, in that she nearer approacheth those parts. For the Moone (as aboue said) hath fiue degrees of latitude: as well to the North, as South: so that when she shall be in the first degree of Cancer, with her greatest latitude Northward; that is, in the head of the Dragon, she shal then be neerer by fiue degrees to the Northerly pole, then the sunne. And in like maner, when she shalbe in the taile of the Dragon (at the entrance and beginning of Capricorne) she shall bee nearer the pole antarcticke by fiue degrees than the sun. Although in the winter the moone should be in the beginning of Capricorn with the Southerly latitude of foure or fiue degrees, yet may she worke and cause more in the change of weather, and shall cause more in Scotland than the sun, in that her power and vertue there is such. But in Brasilia and vnder the antarcticke pole for two causes, the one, in that shee is there of such power, and the other for that in her working she is nearer.

What

of the Circles.

What the longitudes and latitudes of the celestiall Zones are.

The longitude of Zones beginneth from the West, and is extended by the Noonesteede into the East, and from the East againe by the midnight pointe into the West. The motions of the sun in the zodiacke, and Poles of the zodiacke, doe describe the latitude of the zones. For the suns motion or the zodiacke do describe the burnt zone, seeing the sun on the one parte of the zodiacke goeth toward the North vnto the elongation of 23. degrees, 28. minutes, and being by his dayly motion in the beginning of Cancer, doth describe the Tropick of Cancer, which is the bound of the two zones, the burnt zone, and Northerly temperate Zone. And on the other part of the zodiacke doeth the sun goe into the South vnto the same elongation, and being in the beginning of Capricorne, doth likewise describe the Tropicke of Capricorne, which is the bounde of the other twoe zones: in that it distinguisheth the burnt from the southerly temporate zone. And the space also included in these two cyrcles, vsing the middle place, is called the burnt zone, and thus the burnt zone, doth imploy 46. degrees, and 57. minutes.

The Poles of the Zodiacke (which are dayly about the Poles of the worlde) from which they differ 23. degrees, and 28. minutes, and are drawn by the motion of the first mouer, doe describe two cyrcles in the diuers parts of heauen as the Polare cyrcles, which also be the bounds of the zones, that distinguish the twoe temperate from the colde zones. So that the latitude of either colde zone, vnto the

S iij.   poles

poles of the world, is of 23. degrees, and 28. minutes. The other degrees of the semicycle are attributed to the temperate zones; so that either zone containeth 43. degrees, and 3. minutes.

What is the Longitude and Latitude of the earthly Zones.

The longitude of the earthly zones, is like to the longitude of the celestiall, as from the West by the noon sted into the East, and from thence by the midnight pointe againe into the West. And the latitude of them is like to the latitude of the celestial zones: for as the maner of the latitude of the celestiall burnt Zone is vnto the whole cyrcumference; even so is the maner of the earthly burnt zone, vnto the compasse about of the earthly Globe; that is, as 47. degrees is vnto 360. and so likewise conceiue of the others. And that this may plainer appeare, vse the figure following, in which a l h c. is the meridian, or Colure of the solstices, e x l. the Equatoure, a x h. the meridian, s u p. the earthly Globe, s n. the earthly Tropicke of Cancer, k o. the Tropicke of Capricorn e t u. and q p. the arcticke cyrcles. To these answere f r o k bb. and d s m c c. also c ff b a g e c i. the celestiall cyrcles. And what the proportion f d. is, vnto the whole compasse d a k g f the same is (as aboue written) the proportion r s. vnto the whole cyrcumference of the earthly Globe: and on this wise conceiue of the other cyrcles. The letters f d. bee the latitude of the celestiall burnt zone, and r s. of the earthly, d c. and f g. be the latitudes of the temperate zones in heauen, and s t a c r q. of them on earth. The twoe outwarde zones,

of the Circles.

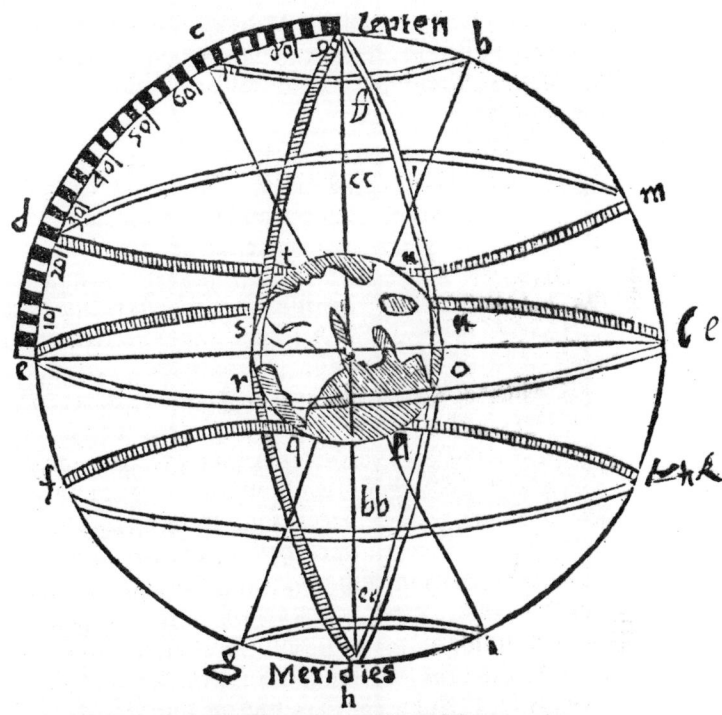

zones, to these here drawne, bee by themselues noted, as well in heauen, as on earth.

Now that wee haue declared with the fiue cyrcles, the latitudes either of the celestiall or terrestriall zones are defined, it shall therefore be necessary to write here of the latitudes of the earthly zones in miles. And that you may readily find the latitude in miles, multiply the degrees by 15. (in that so many Germaine miles, answere to one degree of the great cyrcle in heauen) as the 43. degrees of the burnt zone, being the suns greatest declination, multiplied by the 15. miles, doe produce 705. Germain miles, which

S iij.   is is the latitude of the burning zone. The latitude of either temperate zones, containeth 646. Germain miles almost. And from either Tropicke, vnto the pointes right vnder the poles, doeth the space or distaunce containe 352. Germaine miles.

Where is the beginning and end of euerie Zone, according to latitude, and which places are in which Zones.

The middle of the burning zone is vnder the Equinoctiall line, where either pole is in the Horizon. And both be his bounds (where the eleuation of the pole, aswell Southerly as Northerly is) of 13. degrées, and 28. minutes. For either temperate zone doeth there begin, and streacheth vnto the same place, where the eleuation of the pole is, of 66. degrées, and 30. minutes: which place, is the beginning of the cold zones. By these nowe may a man easily conceiue, which places are in which zone: for if the eleuation of the Pole Northerly, be lesser then 23. degrées, and 28. minutes, this place then is in the burning zone, as the inner Libia, Aethiopia, a part of Arabia Felix, and India. But if the eleuation containeth precisely so many degrées and minutes, the place then is in the bounde of the burnt and temperate zone; as is Siene a city of Aegypt. Further if the eleuation of the Northerly Pole bee greater then 23. degrées, and 28. minutes, yet lesser then 66. degrées, and 30. minutes, this place then is in the temperate zone, as Greece, Italy, Spaine, Germanie, France, England &c. But if the latitude be precisely of 66 degrées, and 30. minutes, the place is in the bounds of the temperate and cold zone, as is almost Lagenlacus

lacus of Suetia. Laſt, if the eleuation of the pole exceedeth 66. degrées, and 30. minutes, the place is in the cold zone beyond which degrées hath Nicolaus Douis a Germaine added a table of Noreway, Gothland, Iſeland, Greenland, Fineland, and Lapeland &c.

How the Zones and Climats doe differ.

The Zone is a space or roome of the earth, from the Weſt into the Eaſt, and from thence by the midnight pointe againe into the Weſt. But the Clymate is a space of the earth, whose beginning is constituted in the Weſt and ende in the Eaſt. A Zone alſo is the space of earth, betwéene two cyrcles equidiſtant, but a Climate is the only space or roome of the habitable earth, contained betwéene two lines equidiſtant.

What the qualities of the Zones are.

TO the celeſtiall Zones are qualities attributed, not formally, but onely vertually; that is, the celeſtial zones are neither cold, hot, nor temperate, but of this named colde, burning, and temperate, through the ſunne, which one whiles comming into this, and another whiles declining into that parts of the worlde, doeth ſend downe his beames to the earth in ſundry maner: as one whiles plum downe right, when the ſun runneth vnder the equinoctiall, and another whiles by a thwart maner, as in the thwart ſphere: which beames (beſides how right angles they make on earth) ſo much the greater heate they cauſe, and how thwarter angles they make, ſo much the weaker heat they procure. So that vnder the Equinoctiall the beames moſt rightly and downe right falling, doe make right angles on the vpper face of the earth, which through the ſame cauſeth a moſt great heat. Alſo the beames faling toward either poles, doe cause thwarter angles, and they make the angles more vneuen or thwarter, and therof the same heat is the lesser. And in the temperate zone (especially in the summer) the beames doe make almost angles falling vnto a rightnes, but in the winter vnto a thwartnes; so that in the same Region is a commodious dwelling. But in the colde zones the angles are caused vnlike or vneuen, & thwartest or slopest, as in the burnt Zone they are rightest and most downward: in somuch that the cold zones euen (as the burnt) are commodious to dwel vnder. For the beames falling and reflected, how much néerer they fal and be togither, somuch the stronger and mightier they moue and cause the heat: as we dayly sée that the sun in the noonstéed being (as in the summer) to cast or streach downe almost perpendicular or down right beames: which beames also are almost reflected into théselues, of which the greatest heat of the day then is caused. And contrariwise, the sun being in the East or west, where ý beames streaching downward and reflected, are scatred and run abroade; the effects be lesser, and the heat much abated and féebled. Euen so the beames in the burnt zone bee perpendicular or plum downright, which reflected into themselues do cause a most great heat. In the temperate zone doe the beames bylitle and litle fall sloper and sloper, of which they cause there a temperate heat. But in the cold zones the beames furthest decline or fall slopest, through which they procure no effect, & of the cosequent cause there a very weake heat.

What the vtilities of the Zones be.

1. The auncient considerers of the stars, haue thus instituted the distribution of the zones for two causes. The one is, that by this reason they might shewe to vs which places of the earth be reasonably habitable, & most commodious to dwell vnder.

2. The other is (as wee learne by experience) that the wits

wits of men, and nature of places by them appeare and are knowne, in that the ayre compassing vs, is a certaine cause of the temperatnes. For the maners and condicions of men (as writeth Galen) doe for the most parte ensue the temperaments of the bodies: yea the nature of trees, plants, hearbs, and beasts do like ensue the temperament of ayre. Of which that we might bee the surer and certainer of the natures of the foresaid matter, it pleased the ancient to deuide them into fiue zones. Of which (it is wel known) that the bodies of men or people dwelling vnder the burning zone (as the Moores) be shorter of stature, then those people dwelling vnder the temperate zones, wilder, and crueler. Also they bee crafty and subtill of nature, hauing besides wrinkled faces, thick crisped heare on the head, and blacke scorched bodies, and crooked of stature. Also all liuing and cresent things, are found to agree according to the quality of the ayre in that Region. Further the people dwelling vnder the Northerly Parallels or Polare cyrcles (where the places bound of colde and moysture) be white of body, hauing long heare on the head, tall and comely of stature and personage, cold of qualitie, yet in maners or condicions wilde and cruell, through the force of the cold in those places, and agreeing with these is the greatnes of the winter, and the greatnesse of fierce and cruel beasts, and other liuing things there breeding, with a furious people inhabiting, called generally the Scythians. Last, those dwelling vnder the temperate zones, be a gentler and ciuiler people beeing some tawnie (especially toward the South) and others toward the North reasonable white of skin and bodie, being meane of stature, and temperate in nature and quality, and of the same like in condicions and behauiuor, &c. And thus much, for the second part of this Treatise.

FINIS.

The Table of all the speciall and seuerall points handled in this Booke.

OF the Rudiments of the Sphere, of Heauen, of the Stars, of the Orbs of the Stars, and of the Earth. Folio 1
What a Sphere is. 2
What the world is, and into how many parts the same is deuided, with the motion of the celestiall Orbs. 8
What the Stars are, and that as to the motion of their Orbes, they are caried about. 11
That Heauen is drawne round. 13
That there are but eight celestiall Orbs, that may be seen. 14
A generall figure declaring the number, disposition, and order of the Celestiall Spheres aboute the Globe of the earth. 16
That there are two first motions of the celestiall Orbs. 17
That there are two kindes of Starres, the fixed and the Planets. 19
Of the celestiall images, and of their diuers names, being in number 48. 21
The 12. signes of the Zodiacke. 22
Of the Southerly. 23
Of the Planets. 25
That Heauen hath a rounde fourme, and is carried circularly. 29
That the water and earth are round bodies, and by a mutuall embracing doe make one body and one hollowe vpper face. 31
An Instrument, by which the roundnesse of the earth (according to latitude) may bee prooued, and all those may easily bee shewed which are taught of the dayes Artificiall. 33
That the water hath a like swelling, and runneth ronnd. 36

That

The table.

That the earth emploieth the middle place of the world, and is the Center of the whole. 42
If the earth be not in the middle of the world, then of necessity it must possesse some of the standings described in the figure there demonstrated, 46
That the earth abideth fixed and vnmoueable in the middle of the world. 49
The phisicke reasons. 50
That the earth compared vnto heauen is as a point. 52
To finde the compasse of the earth, and by it the Dyameter. 55

The second Part.

What the summe of the second part is. 59
That the Sphere of the world is either right or thwart. 90
That the Circles of the Sphere, be some greater, some lesse, with the number of the Circles. 92
The description, names, and vtilities of the Equinoctiall. 62
That this worthy Circle hath diuers names. 71
What are the offices of the Equinoctiall. 73
What are the Northerly images in respect of the Equinoctiall. 78
The description, names, and offices of the Zodiacke, and Eccliptickeline, or way of the Sun. 98
What are the names of this Circle. 100
What is the cause of the thwartnesse of the Zodiacke. 109
Of the Eccliptickeline, or way of the Sun. 113
What the latitude of a Planet is, after two destinctions. 115
What is the longitude of a Star, & where he beginneth. 116
What are the vses and vtilities of the Zodiacke and Eccliptickeline. 124
The description, names, and offices of the Colures. 125
What the offices and vtilities of the Colures are. 133
The description, names, and offices of the meridian Circles, and Horizon. 135

What

The table.

What are the offices and vtilities of the meridian.	144
A Table of the Suns declarations, &c.	155
The common way of measuring of places with their spaces, by the rules of longitude and latitude.	167
What is to bee done if places differ in the longitudes.	169
Other briefe examples.	171
The finding of the distances of places or Citties, in a more easier manner.	173
The first rule.	174
An Example.	175
Another.	Ibid
Another.	Ibid
Another.	Ibid
Another.	179
Another.	Ibid
Another.	Ibid
Another.	177
The second Rule.	Ibid
Another excellent Table, &c.	178
An example of the vse of this Table.	179
The second Rule.	180
An Example.	Ibid
Another.	183
Another.	Ibid
Another.	184
Another.	Ibid
An easier working.	Ibid
An Example.	185
Another.	Ibid
Another.	186
If of two places &c.	Ibid.
A third rule.	190
An example of the third rule.	Ibid.
Another.	193
Another.	124
Another.	196

An

The table.

An easier working and lesse curious.	197
An Example.	198
Another.	Ibid
Another.	199
Another.	Ibid
A demonstration of the third rule.	200
The declaration of the first rule.	201
The declaration of the second rule.	Ibid.
The declaration of the third rule.	202
The definition, appellations, diuision, and offices or vtilities of the Horizon.	204
The appellations and diuers names of the Horizon.	205
The offices or vtilities of the Horizon.	215
Of the verticall Circles.	217
The Circles of the Altitude.	219
The houre Circles.	221
The Circles deuiding the twelue houses of heauen.	226
The Circle of position.	230
The definitions, names, and vtilities of the foure lesser Circles.	231
Which Circles are called the Tropicks.	235
Why they are called Tropicks.	Ibid.
The offices or vtilities of the foure lesser Circles.	245
The descriptions, names, qualities &c.	247
What the longitude & latitude of the celestial zones are.	261
What is the longitude and latitude of the earthly Zones.	263
Where the beginning and end of euery Zone, according to latitude, and which places are in which Zones.	264
How the Zones and Clymates doe differ.	267
What are the qualities of the Zones.	Ibid
What be the vtilities of the Zones.	268

FINIS.

QB
144
H54
1973

AUG 15 1975